THE
STRUGGLE
FOR SUSTAINABILITY
IN RURAL CHINA

THE
STRUGGLE
FOR SUSTAINABILITY
IN RURAL CHINA

Environmental Values and Civil Society

BRYAN TILT

Columbia University Press   *New York*

Columbia University Press
Publishers Since 1893
New York   Chichester, West Sussex
Copyright © 2010 Columbia University Press
All rights reserved

All photographs by author

Library of Congress Cataloging-in-Publication Data
Tilt, Bryan, 1974–
The struggle for sustainability in rural China : environmental values
and civil society / Bryan Tilt.
p.   cm.
Includes bibliographical references and index.
ISBN 978-0-231-15000-2 (cloth : alk. paper)—ISBN 978-0-231-15001-9
(pbk : alk. paper)—ISBN 978-0-231-52080-5 (e-book)
1. Economic development—Environmental aspects—China.
2. Environmental policy—China.   3. Sustainable development—
China.   I. Title.

HC430.E5T55   2009
338.951′07—dc22

2009017783

Columbia University Press books are printed
on permanent and durable acid-free paper.

This book is printed on paper with recycled content.
Printed in the United States of America

CONTENTS

FIGURES

TABLES

PREFACE

THIS BOOK is the result of a decade of research on economic development and environmental degradation in the Chinese countryside. It is based on three periods of ethnographic fieldwork between 2001 and 2006, totaling about seven months of residence in Futian, a township in the southwestern province of Sichuan. I employed a combination of methodological approaches, including approximately 100 semistructured interviews and 150 survey questionnaires with government officials, industrial workers, farmers, and scientists and bureaucrats in the State Environmental Protection Administration. I also relied on participant observation, living for much of my time in the township with Li Jiejie, or "Elder Sister Li," a middle-aged ethnic Shuitian woman from a large and well-connected clan.

Li Jiejie lived in the township's open district (*kaifa qu*) in a modern-style cement house that doubled as the family-planning office, which was overseen by her estranged husband, a cadre in the township government. She operated a small retail store (*xiaomaibu*) out of the front of her house, selling basic household goods and personal items such as laundry soap, chewing gum, and cigarettes. For extra money, she rented my wife and me several rooms and a bathroom with a squatter toilet. Interacting with Li Jiejie, her family, and other villagers on a daily basis allowed me to participate in the ebb and flow of life in the township, to ask the probing and sometimes awkward questions common to anthropologists, and to solicit more detail about the subjects I was studying. I do not intend to portray the township as representative of China as a whole; rather, my goal is to take a close look at how the recent processes of economic development, industrial pollution, and the politics of environmental regulation have played out in this community, and to elucidate what consequences these processes have wrought in the lives of villagers.

Quite a few details in the story of this book relate to China's national environmental-oversight bureaucracy, which, like many things in the reform era, is a moving target. The institutional structure of environmental protection has evolved considerably over the years. From the 1970s through the 1990s, the key agency was the National Environmental Protection Administration. In 1998, the agency was upgraded to the State Environmental Protection Administration. Most recently, in March 2008, the National People's Congress voted to elevate the agency to full ministerial status, renaming it the Ministry of Environmental Protection. Since most of my research was conducted during the SEPA years, I refer to SEPA throughout this book in order to maintain consistent usage. Nevertheless, I want the reader to be aware of the fact that today's MEP is the successor of these previous institutions, similar in its charge but carrying more bureaucratic authority. It remains an open question whether this new administrative structure will help China strike a better balance between economic development and environmental protection.

THE POLITICS OF ENVIRONMENTAL RESEARCH IN CHINA

It has become easier for foreign researchers to work in the People's Republic of China as the nation deepens its political and economic ties with the international community, but it is still by no means a simple task. Acquiring the necessary research permits and credentials is the primary official hurdle. Much of the research for this book was conducted under the auspices of the School of Economics at Sichuan University, located in the city of Chengdu, where I have been fortunate to find many willing and capable scholars with whom to collaborate. But in China, support from above is no guarantee that local officials will cooperate. Before setting up residence in Futian in 2002, for example, my colleagues and I endured scrutiny from the Panzhihua Cultural Office, the Panzhihua Commission for Minority Affairs, and the mayor and Communist Party secretary of Futian Township, each of whom reviewed our research protocol carefully before stamping our paperwork in the official red ink that has been used in civil service offices in China for centuries. The vice general secretary of the Communist Party for Panzhihua City was the highest-ranking official to approve this study. I will not soon forget the trepidation I felt as I waited for him to review my documents, staring anxiously at the door to his office, which bore dual slogans written in beautiful calligraphy: one Maoist ("Serve and People!"), and one Dengist ("Liberate Your Thinking, Seek Truth from Facts!").

On a practical level, language is another barrier to working in China. It is my view that foreigners never become completely proficient in Mandarin Chinese, which requires the memorization of thousands of ideographic characters and the ability to distinguish and reproduce the four tones of the spoken language. Rather, foreigners become adept, to varying degrees, at getting by in the language. In my case, my years of studying *putonghua*, or standard Mandarin, in courses and from textbooks proved only partially useful in Sichuan, where peasants speak a dialect called Sichuanhua. Speakers of Sichuanhua, whose vowels and consonants depart considerably from *putonghua*, also use a lexicon that is impossible to learn by any other means than immersion. During my first experience interviewing villagers in 2001, for example, informants responded to my questions with a patois completely unintelligible to me. Notebook and pen in hand, I had to turn with a shrug to my companion, a senior anthropologist from the Sichuan Nationalities Research Institute, who handily translated their responses into *putonghua* for me.

My position in the community as a researcher was perhaps an even thornier issue. As a foreigner, arriving on the scene in rural China and asking probing questions about the economic and environmental consequences of local factories likely made me seem presumptuous to many villagers. Having emerged from decades of socialist rule, during which candor with outsiders could have unforeseen negative consequences, villagers can be quite closed-lipped even on the most benign of topics; on the politically charged subject of pollution, they can be positively mum. In my experience, the ethnographic method provides its own partial solution to this problem; by continued presence in the community, I gradually gained the trust of most local villagers and cadres, who slowly came to the conclusion that I was relatively harmless and allowed me to go about my work unimpeded.

Nevertheless, villagers often suspected that I was aligned with one or more interest groups in the community. People who routinely saw me ascending the winding dirt road to the industrial compound to interview factory workers, for example, assumed that I was serving as some sort of economic advisor to the factories. Industrial workers, meanwhile, whose livelihoods were precarious because of monitoring and inspection activities conducted by the district Environmental Protection Bureau, often assumed that I had an environmental agenda. On one awkward occasion, I even had to refuse a ride to the industrial compound in the EPB Jeep, opting instead for a steep hike in order not to be seen arriving at the factories in the company of regulatory officials.

Early in my field research, it was made clear to me that my project entailed a degree of risk for the people with whom I worked. One afternoon, as I interviewed a woman in Li Jiejie's store, she appeared somewhat uncomfortable about answering the questions related to her perceptions of pollution from local factories. At that moment, several cadres drove past the storefront in the township government Jeep and the woman, looking quite panicked, quickly ducked behind the counter to avoid being seen talking with me. In time, incidents like these decreased in frequency, but I always remained mindful that there were real-world consequences to my work for many of my informants. Many people preferred to remain anonymous, and I have respected their wishes, and the mandates of my university agreement regarding the treatment of human research subjects, by using pseudonyms or referring only to their surnames. For government officials or others involved in public service, whose identities are a matter of public record, I use real names and administrative titles, remembering that they were operating both as institutional representatives and as individuals with unique personalities and actions. With this book, I hope that I have painted a picture of Futian that local people would recognize, appreciate, and find valid and insightful.

ACKNOWLEDGMENTS

I have accumulated many debts during the process of writing this book. I would first like to acknowledge with gratitude the people of Futian Township, who not only participated in my research project but also offered their friendship, their homes, and their food to me during my fieldwork. In the local township government, Wang Chunyang and Zhen Dehua, who both served terms as Communist Party Secretary, allowed me to conduct research in their jurisdiction that carried potentially serious political consequences. Zhang Yunfu and Li Fuming, the township's mayor and vice mayor, respectively, also provided key research assistance. Li Jiejie and various members of her extended family became true friends and confidantes.

I am intellectually indebted to many colleagues and friends, whose ideas and input have shaped this project over the years. Most significantly, Stevan Harrell, my doctoral advisor and friend, has had a hand in this work from the beginning. His reputation among scholars and government officials in

Sichuan opened many of the doors that made this research possible, and his intellectual input has been invaluable.

I have also benefited greatly from the knowledge, criticism, and support of others who have read part or all of this book along the way, including Eugene N. Anderson, Marco Clark, Li Yongxiang, Sally Liu, David McMurray, Bettina Shell-Duncan, K. Sivaramakrishnan, Courtland Smith, Clark Sorenson, and Susan Whiting. I am also grateful for the comments of several anonymous reviewers for Columbia University Press. Loretta Wardrip and Lindsay Nansel provided valuable assistance with the illustrations.

Any serious research project is by necessity a collaborative effort; in China, where foreign researchers are confronted with myriad bureaucratic obstacles, this is doubly true. I owe a particular debt of gratitude to the many Chinese scholars who saw value in this project and helped me see it to fruition. Li Xingxing of the Sichuan Provincial Institute for Minority Nationalities Studies accompanied me in 2001 on my first visit to Futian. Possessed of a keen knowledge of the diverse ethnic minorities of southwest China, he served simultaneously as a skillful ethnographer, impromptu translator, and liaison with local government officials. I conducted some of my fieldwork while in residence at Sichuan University as a visiting scholar. Yan Shijing and Chen Bing, in the Office of Foreign Affairs, helped smooth the way for me not only to come to Sichuan University but also to reside locally in Futian and conduct proper ethnographic fieldwork, an endeavor they thought somewhat strange. Zhu Fangming and Xiao Pichu of the School of Economics at Sichuan University lent their expert knowledge of rural industry to the project. Xiao Peichu, who cut his academic teeth along with me during the course of this project, logged many hours of fieldwork and helped to build in-roads with local government officials. Over the years, he has become a trustworthy collaborator and good friend.

I would also like to acknowledge the financial support of many generous institutions. Early support was provided by the Department of Anthropology and the East Asia Center at the University of Washington. A National Science Foundation Dissertation Improvement Grant (NSF-DIG 0210178), jointly funded by the Division of Cultural Anthropology and the International Division, supported the field research upon which much of this book is based. During 2007 and 2008, I held a fellowship at the Center for the Humanities at Oregon State University, which supported further research and writing and facilitated stimulating interactions with colleagues from various disciplines. I have also been buoyed along over the past few years by a generous research grant from the College of Liberal Arts at Oregon

State University, and by David McMurray, the chair of the Anthropology Department, who allowed me to keep this project at the top of my priority list, sometimes at the expense of other duties.

I could not have completed this project without the personal support of friends and family. I owe the deepest debt of gratitude to my wife, Jenna Tilt, who conducted fieldwork in China with me twice under trying circumstances and who took many of the photographs that illustrate this book. She is also a true intellectual partner in this project, since she shared with me many of the discoveries made during the research process. My children, Avery and Miriam, are, as always, my greatest points of inspiration. I also acknowledge the constant support of my parents and brother.

Part of chapter 6 is based on materials from my article "The Political Ecology of Pollution Enforcement in China: A Case from Sichuan's Rural Industrial Sector," published in *The China Quarterly* 192 (2007). Part of chapter 7 is based on materials previously published in an article entitled "Industry, Pollution, and Environmental Enforcement in Rural China: Implications for Sustainable Development," in *Urban Anthropology and Studies of Cultural Systems and World Economic Development* 36, no. 1–2 (2007). At Columbia University Press, I am grateful for Anne Routon's tireless efforts on my behalf and for Michael Haskell's careful editing and production work. I am responsible for any errors or shortcomings that remain in the book.

THE
STRUGGLE
FOR SUSTAINABILITY
IN RURAL CHINA

1

ENVIRONMENTAL VALUES, CIVIL SOCIETY, AND SUSTAINABILITY IN POST-REFORM CHINA

O N A summer evening in 2006, Mr. Zhen Dehua, the Communist Party secretary of Futian Township, hosted a modest banquet of roasted rabbit and chicken soup in honor of my return, along with a Chinese economist colleague, to conduct research in the area.[1] The monsoon rains had not yet arrived, and the air in the makeshift banquet hall, which normally served as a meeting room in the township-government office building, was hot and stifling. Party Secretary Zhen led the guests, who also included the mayor, the vice mayor, and the head of the township industrial-development office, in a series of toasts. Raising his glass of *baijiu*, the distilled liquor that is ubiquitous in China, he eloquently quoted a well-known series of couplets attributed to Deng Xiaoping, the architect of China's liberal economic reforms of the past three decades:

Wu liang, bu wen;
Wu gong, bu fu;
Wu shang, bu huo.

[Without grain, there is no stability;
Without industry, there is no wealth;
Without commerce, there is no vitality.]

Party Secretary Zhen's words, consisting of a classical structure of three balanced lines with four characters each, were meant partly as an homage to one of China's great and visionary leaders and partly as a eulogy for the golden age of industrial development in Futian, which now seemed to lie inexorably in the past. On previous visits to the township, I had noted one of the jarring ironies of life in this mountainous corner of Sichuan province: just when the terraced paddies, mud-walled houses, and grazing water buffalo had convinced me that I was in the "genuine" Chinese countryside, I encountered yet another hilltop of rusted factories belching black, acrid smoke into the air. In China, city and countryside are becoming harder to distinguish. But in the summer of 2006, there was no smoke and no sulfurous smell in the air in Futian. For three years, the township's factories, which included a zinc smelter, a coking plant, and a coal-washing plant, had been closed for noncompliance with nationally mandated pollution standards. Piles of raw coal, ore slag, and metal scraps were all that remained in the dormant factory compound on the hilltop on the north edge of the township. The future of industry in Futian, and the rising standards of living it supported, were now uncertain. Mr. Zhen and the other township cadres were visibly concerned.

Understanding the economic consequences of this turn of events requires some grasp of China's recent development path. The socialist revolution that ushered the Chinese Communist Party into power in 1949 was rooted in the Marxist principles of collective ownership of the means of production. When Deng Xiaoping assumed power in 1978, he announced the beginning of a suite of liberal economic policies known as "Reform and Opening [*gaige kaifang*]." Thirty years of gradual reform have brought sweeping social and economic changes to the People's Republic of China, including the return of smallholder agriculture under the Household Responsibility System, the privatization of industrial production, greater integration into the world economy, and the rise of an urban consumer class. China's gross domestic product has grown nearly 10 percent per year over that time period, and its economy is expected to be the largest in the world, surpassing the United States, within the next two decades.

Visitors to China often remark half-jokingly that the national bird is the crane: not the white, majestic red-crowned bird native to the marshes of Manchuria but the six-story construction crane that can be seen erecting the steel, glass, and cement high-rises of China's modern cities. These are perhaps the most recognizable symbols of China's ascendant place in the global economy. Equally dramatic, however, are the changes that have swept through rural China over the past several decades, transforming the lives of the nation's 800 million peasants. This book deals in particular with the most dynamic and perhaps most peculiar aspect of economic development in contemporary China: rural industrialization.

Within a few short years of his ascent to power, Deng Xiaoping and his political allies began experimenting with decollectivized agriculture in selected villages in Sichuan and Anhui provinces. By the early 1980s, agricultural communes and production brigades were systematically dismantled, and peasants throughout China became subject to the Household Responsibility System, under which they were granted long-term land leases and the opportunity to sell excess grain and other crops in the nation's burgeoning markets. As any economist might predict, this de facto privatization of land created a powerful incentive system for farmers, whose redoubled efforts produced crop yields never before seen in China.

By the 1980s, local governments throughout rural China began investing in relatively low-tech, labor-intensive factories. These served two important purposes: they absorbed some of the surplus labor in the countryside, which had been created by more efficient agricultural practices, and they provided township and village governments with a much-needed tax base. The Chinese countryside was dotted with literally millions of these metal smelters, brick furnaces, cement plants, and fertilizer factories. These factories, which the government dubbed "township and village enterprises," provided a new way for local cadres to absorb surplus labor, secure steady operating revenues for the township or village, and boost household incomes in rural areas. By the early part of this decade, China boasted millions of rural factories employing more than 135 million workers and accounting for one-third of the nation's GDP (China TVE Yearbook 2004). No single development policy has proven so successful at lifting millions of rural people out of dire poverty.

However, Chinese leaders have pursued this path to industrial development with what the historian Judith Shapiro has aptly called "utopian urgency," and for the most part they have done so with little regard for the environmental consequences (Shapiro 2001). The black smoke, sulfurous air, and contaminated water that became so familiar to me during my fieldwork in Futian are facts of life for villagers in all but the most remote corners of China. In fact, rural factories are responsible for up to two-thirds of China's total air- and water-pollution burden (China Industrial Development Report 2001). Rural industry has thus become an emblem of the difficult questions facing contemporary China about how to balance the immediate and acute need for economic development with the long-term goal of environmental sustainability.

The severity and scale of China's environmental problems, of which rural industrialization is but one cause, has garnered the attention of the global media. One recent news headline declares, "Pollution Turns China Village Into Cancer Cluster" (Liu 2005). In 2007, the New York Times ran a special series of feature articles on China entitled "Choking on Growth." Much of this attention is warranted. Water quality in most major rivers is so terrible that half of the river sections that undergo routine monitoring earn a class 5

rating, the lowest on the nation's environmental-quality standards for surface water. The World Bank reported in 2001 that sixteen of the world's
twenty most polluted cities were in China (Economy 2004). In major cities
such as Beijing, annual average concentrations of particulate matter frequently surpass health and safety thresholds recommended by the World
Health Organization for a single daily average (Smil 2004:17). Sulfur dioxide
emissions, mostly from coal-fired industrial plants, cause acid rain across
much of the country. Meanwhile, coal continues to supply two-thirds of
China's energy demand, and coal reserves are sufficient to burn at current
levels for most of the coming century. The environment may well be the first
and most meddlesome check on China's meteoric economic growth.

As an anthropologist, I have focused my research for most of the last
decade on understanding the ecological and health risks from industrial
pollution in rural China. When I present my research to other scholars and
to the general public, I hear two common responses. Over time, these critical responses, along with continued interaction with township residents,
government officials, factory workers, and environmental regulators in
China, have influenced my thinking to the point of shaping the conceptual
framework of this book. The first response is that Chinese people likely
don't care about pollution because they are focused on the immediate need
for economic development. The second is that even if people did care about
pollution, public opinion doesn't matter very much in the context of an
authoritarian, single-party government that consistently emphasizes economic development over environmental protection.

In this book, I use these two critical responses as points of departure for
thinking about the processes and consequences of China's industrial development path and the prospects for achieving environmental sustainability.
Pollution control and sustainable development in China have become the
subjects of considerable academic scrutiny, primarily from a macroscale
policy standpoint (Day 2005; Economy 2004; Smil 2004; Ma and Ortolano
2000). While I have personally found much of this research to be quite insightful, I am also continually surprised at the lack of community-level ethnographic studies on China's environment, which is perhaps the most crucial issue facing this nation of 1.3 billion people. Scholarly treatments of
China's environment often quickly devolve into faceless recitations of dismal statistics. In my view, such an important topic, and one germane to the
daily lives of so many people, cannot be studied solely from macrolevel
analyses or reviews of existing data and literature. Rather, what is required
is an in-depth engagement with the people who are at the center of such
dramatic social, economic, and environmental changes. In focusing ethnographically on Futian Township, my aim is to explore the lived experiences

of villagers as they cope with severe changes to their local environment, natural resources, and health and welfare. In what follows, I discuss the main themes of this book and describe how this ethnographic study can contribute to a fuller picture of China's current environmental problems and the search for solutions.

ENVIRONMENTAL VALUES
IN CONTEMPORARY CHINA

As will become clear in the course of this book, there are obvious technical and institutional reasons for China's enormous pollution problems, including a lack of capital to invest in environmental-mitigation technologies, funding and manpower shortages in the national environmental-oversight bureaucracy, and heavy fiscal interdependence between local governments and local factories. But most experts on China's environment agree that enforcement of existing laws and regulations is the key missing link in the effort to improve environmental quality (Ma and Ortolano 2000). Enforcement requires a strong regulatory framework and an environmental bureaucracy capable of implementing that framework. But it also requires public awareness of environmental problems and public support for doing something about them.

While China's rich philosophical traditions, including Taoism, Buddhism, Confucianism, and folk religion, emphasize "harmony between people and nature [*tian ren heyi*]," twentieth-century Chinese history, with such spasmodic campaigns as the Cultural Revolution, effectively erased much of this environmental ethic from the collective consciousness. In the current literature on environmental protection in China, environmental values and even basic environmental awareness are often assumed to be absent. Peasants in particular are depicted as too poor, too uneducated, and too concerned with the exigencies of making a meager living to worry about environmental problems (Wheeler, Wang, and Dasgupta 2003; Lo and Leung 2000; Edmonds 1998). The prevailing scholarly view thus envisions a Chinese peasantry fatalistically breathing polluted air and drinking contaminated water in the name of national economic development: hence the common refrain I often hear from colleagues, students, and the general public about Chinese people's low level of environmental consciousness.

In fact, this idea is drawn from and supported by much of the scholarly work on global environmental values. Perhaps the most influential work on the emergence of environmental awareness is the "postmaterialism thesis," most closely associated with the World Values Survey conducted by Ronald

Inglehart and colleagues at the University of Michigan on an ongoing basis since the early 1980s. The survey has collected data from dozens of countries on all six inhabited continents in the world. In brief, the postmaterialism thesis suggests that economic wealth causes a fundamental shift in human values; as societies undergo the transition to industrial development and modernity, their citizens begin to concern themselves with needs and wants beyond the material, including gender equity, quality of life, happiness, self-expression and spiritual fulfillment (Inglehart 1997, 1995). By extension, the argument holds that environmental concern can be found primarily among wealthier nations and individuals, while the poor cannot afford the "luxury" of worrying about environmental protection (Dunlap and Mertig 1997; Dunlap, Gallup, and Gallup 1993b). The economist Juan Martinez-Alier (1995) felicitously called this being "too poor to be green."

More recently, scholars have begun to focus specifically on the environmental values, attitudes, and actions of people in developing countries. The 1992 Health of the Planet survey, conducted by Gallup International Institute, found that citizens of rich and poor countries alike were extremely concerned about the environment (Brechin 1999; Dunlap, Gallup, and Gallup 1993b; Dunlap, Gallup, and Gallup 1993a). Similarly, the Global Environmental Survey, an ambitious attempt to measure the attitudes and behaviors of citizens from multiple countries regarding the environment and sustainable development, provides further evidence of growing environmental concern in the developing world (Ester et al. 2002). In fact, it now appears that two kinds of environmental consciousness may be emerging: one, called "full-stomach environmentalism," is common in the global North and is the product of shifting values because of affluence; and the other, called "empty-belly environmentalism," is common in the global South as people experience directly environmental problems, such as pollution, that continue to be acute in the developing world and are often the result of First World countries externalizing their own environmentally damaging behaviors (Martinez-Alier 2002; Guha and Martinez-Alier 1997).[2]

This trend is evident in the World Values Survey data itself: while China is considered a lower-middle income country by the World Bank, it also ranks in the upper quartile in terms of public support for environmental protection, along with some much wealthier and decidedly postmaterialist nations such as Sweden, Denmark, the Netherlands, and Norway (Inglehart 1995). This seems incongruous until one considers the fact that China also has the worst urban air-pollution rating of any country surveyed. Exposure to the ill effects of industrial development can and does result in heightened environmental awareness, even in cases of acute economic need.

While little detailed information is available about the environmental consciousness of Chinese citizens, several recent national studies, combined with smaller regional studies, have produced limited and mixed results. Urban and rural residents both appear increasingly concerned about environmental problems such as air and water pollution, deforestation, and acid rain, but their concerns are balanced against other pressing issues such as economic growth, employment, population control, and public order (Lee 2005; Xi and Qinghua 1999). Environmental issues in China appear to be a high priority for urban, educated Chinese but do not seem to hold the salience necessary to produce an organized environmental movement (Stalley and Yang 2006). Noting the complex nature of global and local environmental values and their influence on policy and behavior, the sociologist Steven Brechin has called for renewed social-science inquiry into environmental values as they are situated in specific places and contexts:

> Environmentalism is most likely a complex social phenomenon, a mixture of social perceptions, local histories, and environmental realities, international relationships and influences, and unique cultural and structural features of particular countries and regions. . . . To better explore global environmentalism, we ideally need to generate more systematic research on the social bases of environmentalism in every country and region of the world, or a reasonable sample thereof.
>
> (BRECHIN 1999:807)

Nowhere is this call for research more pertinent, and more timely, than in China, where environmental problems seem increasingly intractable. By taking a fine-grained ethnographic approach to the study of environmental contamination in Futian Township, I hope to also deepen our understanding of environmental values in reform-era China.

While the prevailing methodological tool of standardized surveys can play an important role in identifying trends such as the rise of global environmental consciousness, the method presents us with two difficult challenges. First, most people experience environmental problems such as pollution in profoundly local ways; they may notice the effects on their personal health, their livelihoods, or their food safety. By aggregating data at the national or regional level, large-scale surveys mask a great deal of the richness in how people actually think about and interact with their environments. Second, international surveys often have trouble framing the issues in ways that are culturally, politically, and economically salient for people. In a recent book on the historical and cultural aspects of environmental values in China and

Taiwan, the anthropologist Robert Weller points out, for example, that his informants "lack environmental consciousness only in the sense that they are not concerned with the same issues as national and global elites, or as people who write questionnaires about values" (Weller 2006:157).

A complete understanding of environmental values, as well as an appreciation of how values shape actions and behaviors, requires in-depth studies of specific people and places. As this book shows in detail, villagers in Futian Township were acutely aware of and concerned about pollution from local factories. Despite the fact that rural industry has played a key role in the development strategy of the township, most local people recognized—and indeed experienced daily—the ill effects of pollution on their health, on the quality of the agrarian ecosystem, and on their economic livelihoods.

CIVIL SOCIETY AND THE ENVIRONMENT

The second common response I hear about my work is that environmental attitudes matter little in China, where the authoritarian governmental regime routinely sacrifices environmental quality for economic gain. In essence, this point is about the role of civil society in addressing China's environmental ills. In using the term "civil society," I am referring to an intermediate realm between the family and the state characterized by collective action around shared values, interests, and goals (Hefner 1998; Seligman 1992). As the Chinese Communist Party has liberalized the nation's economy over the past three decades, it has also gradually reduced the scope of its administrative power, increasing the space within which civil-society organizations may operate (Weller 2005). Yet the Chinese case presents us with some special challenges in thinking about civil society in regards to the environment, since there is little precedent for the protection of individual rights and the single-party political system severely constrains collective action. In short, a set of empirical questions arises: Where do civil-society strategies come from? What cultural values, historical precedents, and legal statutes underpin them? In what social and political contexts are they deployed? How do various facets of civil society in China engage with the nation's seemingly intractable environmental problems?

As I illustrate in this book, China's rapidly deteriorating environment is not merely a material problem but a sociopolitical problem as well. This means that the government and its citizens are confronting new questions about the role of public participation in environmental decision making, the uneven distribution of environmental risks, and transparency in the environmental oversight process. The situation is complicated by a political atmosphere in which decision-making authority and control of information is concentrated in the hands of the Chinese Communist Party. Environmentally oriented citizens' groups face an array of problems in China, where they lack "both the opportunity and the immediate urgency to openly confront the central government" (Ho 2001: 897).

While we should exercise caution in assigning an undue degree of autonomy and power to nonstate actors in Chinese environmental politics, the role of civil society cannot be entirely ignored, either. Registered environmental NGOs are subject to a dizzying set of controls and regulations from China's State Council and Central Committee, but informal citizens' groups in fact often find creative ways to accomplish their goals within this complex, authoritarian system. Their strategies for addressing environmental concerns include lawsuits, petitions, and sometimes even open protest (Jing 2000). More and more empirical evidence points to the rise of "rightful resistance" among citizens' groups who wish to encourage a new direction in environmental politics and processes. The political scientists Kevin O'Brien and Lianjiang Li (2006:2) describe rightful resistance as:

> A form of popular contention that operates near the boundary of authorized channels, employs the rhetoric and commitments of the powerful to curb the exercise of power, hinges on locating and exploiting divisions within the state, and relies on mobilizing support from the wider public. In particular, rightful resistance entails the innovative use of laws, policies, and other officially promoted values to defy disloyal political and economic elites.

It is important to distinguish rightful resistance from open rebellion against existing power structures, which in contemporary China can prove politically ineffectual and downright disastrous. Rightful resistance must remain within the bounds of legality set by the government and, if it is to be successful, must invoke some of the ideas and discourses of the government itself. There are some signs that these strategies of rightful resistance are beginning to bear fruit. For example, the discourse of environmental protection within the central government is gaining momentum; in 2005 China's State Environmental Protection Administration, with the support of Premier Wen Jiabao, halted thirty major industrial projects for failure to properly conduct environmental-impact analyses. Environmental protests are on the rise throughout China, in part because the central government is growing more tolerant of these movements, provided they remain local in their scope of interest and relatively small in scale (Oi 2001). Information technologies such as the Internet help to support the dissemination of information and the organization of civil society around shared interests such as the environment (Yang 2005:58–59).

In a sense, these new developments in environmental politics encourage us to begin thinking about the Chinese state in a new way. It is tempting to see the Chinese government, from the halls of power in Beijing to the villagers' committee meetings held in even the most peripheral locations, as a singular, monolithic entity. But in fact, the dynamic tension between center and periphery is a common theme in much of Chinese political history. This tension has, if anything, deepened during the reform period as the central government retreats from some of its traditional duties as a result of liberal economic reforms. Where the government once controlled the production and consumption of goods, supplied its citizens with social-welfare programs, and guaranteed lifetime employment, it is now conceding ground to a growing private sector. The retreat of the central government from public and private life is opening up space for new kinds of social action in response to environmental problems. One aim of this book, therefore, is to explore the strategies used by Chinese citizens to gain redress from pollution, and to follow the environmental oversight process in Futian, tracing the ins and outs of pollution enforcement as they unfold on the ground.

THE STRUGGLE FOR SUSTAINABLE DEVELOPMENT

These critical responses get us to an important line of inquiry about sustainable development in contemporary China. In the arena of international development, few recent concepts hold the salience and widespread public approval of "sustainable development." The darling of environmental activists and multilateral development agencies alike (including the World Bank and the United Nations Development Program), the concept is quite powerful in its simple assertion that economic growth should be undertaken within ecological limits. Sustainable development has become a household phrase and a cottage industry since the UN Commission on Environment and Development produced its seminal report, *Our Common Future*, in 1987.

China was a high-profile signatory to Agenda 21, a policy document that emerged from the UN Earth Summit in Rio de Janeiro, Brazil, in 1992 and emphasized the need to protect natural resources for use by future generations. In fact, with more than 1.3 billion people and an overheated economy that has become the world's manufacturing hub, China is undeniably a key player in international environmental politics. The rise of a new consumer class in China, with its disposable income and its growing desire for durable consumer goods such as automobiles, raises the specter of what might become of a world in which vast numbers of people attain the living standards of the West. What would be the demands on energy, waste disposal, and housing? One paragraph of Agenda 21 muses about the likely effects, no longer hypothetical, of "a refrigerator in every Chinese household."

Many scholars are skeptical of China's commitment to sustainable development in policy and practice, and the seemingly intractable state of the nation's environmental problems certainly warrants such skepticism. Mao Zedong himself once famously declared that "if people in nature want to be free, they will have to use natural sciences to understand nature, to overcome nature and to change nature; only then will they obtain freedom from nature" (Mao 1966:44). His statement encapsulates many of the troubling elements in socialist China's environmental ethos: an epistemological separation of human beings and nature; a penchant for using science in an idealized, uncritical fashion; and a tendency to see the natural world as yet another force, like social class, that must be "struggled" against if the goal of human progress is to be achieved.

Yet contemporary China is undergoing a sea change in its environmental discourse and practices. Its ascendance on the world economic stage and its participation in the key forums regarding sustainable development have generated a domestic environmental agenda with considerable momentum focused on the concept of sustainable development, which state-sponsored publications translate as "*kechixu fazhan*," or "development which can be sustained." In his final speech as head of state before the Sixteenth National Congress in 2002, President Jiang discussed the meaning of *xiaokang* (literally "small comfort," or being well off) within contemporary Chinese society, a topic which was one of the cornerstones of his leadership. That he made this key term part of his address to the NPC is therefore barely worth noting, but his particular redefinition of the term is significant. Jiang laid out what he called "four main goals in establishing a *xiaokang* society [*jianshe xiaokang shehui si da mubiao*]," the last of which, despite its obtuse language, unmistakably underscores the national government's interest in environmental sustainability:

> [China should] continuously strengthen the ability to attain sustainable development [*kechixu fazhan*], better the natural environment, improve efficiency in the use of natural resources, attain harmony between man and nature, push toward the development of the forces of production for all society, a rich life, and the road leading to the development of a better environment and civilized society.
>
> (CHINA TOWNSHIP AND VILLAGE ENTERPRISES 2002:4)

This official discourse is backed by a growing body of environmental law and a burgeoning of the State Environmental Protection Administration, China's environmental oversight bureaucracy, which was promoted to ministerial status in 1998. In my conversations with local and regional environmental-protection officials, some of whom were responsible for monitoring Futian's factories and enforcing pollution standards, I was struck by how often they invoked the concept of sustainable development as they described their job duties and their agencies' objectives. As one official told me, "Once you've reached a certain level of development, then you've got to consider environmental problems." The global discourse of sustainable development is currently undergoing a process of sinicization.

But to say that sustainable development operates as a discourse in China tells us little about exactly how it operates and with what consequences for communities and individuals. In fact, it raises a key question: Even if Chinese leaders agree on sustainable development as a goal, what precisely should be *sustained*? In the wake of factory closures in Futian, for example, township cadres, industrial workers, and villagers were quite divided on whether sustainability meant preserving the quality of the natural environment or retaining important industrial revenue and wages. The enforcement of pollution standards thus became a poignant collision of economic and environmental concerns, in much the same way as it is portrayed in the international literature on sustainable development (Osorio et al. 2005; Parris and Kates 2003).

One contribution of this book is to explore in depth how the global discourse of sustainable development shapes environmental policy and action on the ground in China. This is not merely a theoretical question but one with profound consequences for communities such as Futian, where the trend toward controlling industrial pollution in the name of sustainable development has a direct effect on people's economic livelihoods. I do this by both examining the growing discourse of sustainable development in China and by focusing ethnographically on the processes through which this discourse is translated into action with regards to pollution control.

FUTIAN TOWNSHIP

The liberal reforms of the last thirty years have generated tremendous wealth in China, but they have also created and exacerbated economic inequality. The rural-urban disparity can be truly shocking to the foreign visitor; I have often reflected that I experience less culture shock when I get off the airplane from the west coast of the United States to Beijing than I do when I board a train or a bus and travel from the city into the Chinese countryside. There is also a widening economic disparity between regions as one moves from the populous eastern coast, where Special Economic Zones have fueled China's integration into the global economy, to the interior hinterlands of the country. The southwestern region, which is the setting for this study, has been the ethnic, cultural, and economic frontier of China for many centuries. It is here that the nation's development challenges can be seen in greatest relief.

Situated in the southern tip of the mountainous province of Sichuan, Futian Township was named after its "bountiful fields," an ironic fact when one considers the arid, steep landscape that greets the visitor. It is located on the western edge of Panzhihua Municipality, on the foothills of the Qinghai-Tibet Plateau (see figures 1.1 and 1.2). The region takes its name from the panzhihua tree (*Ceiba pentandra*), a beautiful specimen whose name literally means "climbing branch flower" and whose yellow, orange, or red blossoms herald the coming of springtime.[3] Futian township, which encompasses the four villages of Wuzitian, Jingui, Guantian, and Tangba, comprises mostly one-story mud houses with tiled roofs.[4] In the center of town, which residents refer to as the "open district," or *kaifa qu*, new multistory cement buildings with storefronts line the roadside, and villagers can be seen bustling to the morning market, chatting with one another, or playing mahjong. The tallest building in town is the six-story government office building, an impressive structure with a clean, white-tile façade that was constructed in the late 1990s.

A small stream running through the township feeds into the Jinsha River as it flows from its source high on the Qinghai-Tibet Plateau toward Panzhihua, at which point it becomes known as the Yangtze. The township is situated in a valley that ranges in elevation from 1,000 meters along the streamside, where the land is terraced for growing rice, to more than 1,600 meters where the ground is steep, dry, and covered in scrub brush, Yunnan pine, and eucalyptus. In the distance loom mountain peaks that rise to more than 2,500 meters. Except for the monsoon season, which arrives during the summer months, Futian is typically dry and dusty. Its iron-rich, rust-colored soils are reminiscent of the American Southwest, and its climate is similar, too: summer temperatures can reach forty degrees Celsius, and winters are quite mild. Despite the adage, common throughout the southwest region, that one must "remove a wheelbarrow full of rocks just to grow a mouthful of rice," agricultural yields in Futian are generally sufficient

FIGURE 1.1 Sichuan Province and the study region.

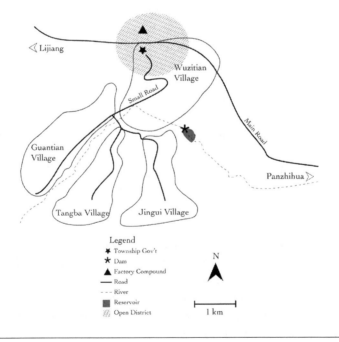

FIGURE 1.2 Futian Township.

to feed the local population and to sell excess grain, vegetables, and meat as far away as Panzhihua.

Yet food security was achieved relatively recently in Futian; many villagers recalled poverty so acute during the Maoist period that they were forced on occasion to eat forage grass along with their animals just to survive. Most families in Futian held between one and five *mu* of land, typically consisting of both irrigated rice paddy and dry fields for planting crops that can be irrigated by carrying water from the stream.[5] The average household income in the township was 3,637 yuan in 2004, slightly above the provincial average and well below the national average. The Engel's coefficient, which is a measure of the percentage of household income spent on food, and a good proxy measure for poverty, was 55.6 percent (Sichuan Statistical Bureau 2005:175). In one of my interviews, a villager aptly summed up the standard of living in the township as "neither rich nor poor" by Sichuan standards (*bu qiong, er bu fu*).

Futian's industrial compound (*gongye yuanqu*) was situated at the northwestern edge of the township, along the main road between Panzhihua to the east and Huaping, a county town across the Yunnan border to the west. A patchwork of low-tech, dilapidated factories, including a zinc smelter, coking plant, and coal-washing plant, the compound was home to about a hundred industrial laborers and their families, many of whom had migrated hundreds of kilometers to live and work in Futian. On a typical day before environmental-protection officials shut down the factories, plumes of black smoke could be seen rising above the industrial compound. The air had an acrid, metallic smell to it, and the water in the stream often ran black with the effluents from the coal-washing plant. Piles of coal and ore slag lay strewn about the factory compound. When it rained, pools of black industrial sludge collected in the ruts and potholes of the road and in villagers' courtyards and gardens.

Futian provides an apt setting for examining the processes and consequences of industrial development because it allows us to follow China's arc from the state- and collectively owned enterprises of the socialist period to the market-oriented, private enterprises of today. It also allows us to explore villagers' attitudes about and coping strategies related to industrial pollution. Futian is also a mixed ethnic community, home to Han Chinese and Yi minority-nationality people. On the ethnic peripheries of China, the problems of economic development are most pronounced, and the achievement of industrial success is most crucial. The imperative for economic development is as strong here as anywhere in China, while the negative consequences of unfettered industrial development have also become impossible to ignore. Villagers faced an uncertain future as they grappled with the costs of industrialization, including damage to crops and livestock, growing disparities in wealth, and the deterioration of health in their community.

ENVIRONMENTAL ANTHROPOLOGY
AND THE BROWN RESEARCH AGENDA

My inquiry owes a great deal to the rich intellectual lineage of ecological and environmental anthropology, which I would like to contextualize briefly, if only to hint at where this book fits within the broader schema. Early scholarly efforts in ecological anthropology—including the pioneering work of Julian Steward on cultural ecology and Roy Rappaport's research on energy flows and feedback loops in human ecological systems—focused on the ways that culture, shaped by environmental conditions, served as an adaptive mechanism for human populations living in various ecosystems. Borrowing many of its explanatory models from the nascent science of ecology, these and other foundational scholars examined the processes through which nature and culture interacted to maintain homeostasis.

The field has come a great distance since the mid-twentieth century and has been influenced along the way by broader trends in the social sciences, including Marxist political economy, poststructuralism, discourse analysis, and critical-development studies. The "new ecological anthropology," as Kottak (1999) has called it, "blends theory and analysis with political awareness and policy concerns" by focusing on specific processes and problems—soil erosion, deforestation, the collapse of fisheries, or toxic contamination. For most contemporary environmental anthropologists, the analytical focus is on the complex structural, economic, political, and cultural processes underlying environmental change.

Under this rubric, I view the current field of environmental anthropology as divisible into two broad strands. One, what I think of as the "green research agenda," focuses on modern conservation policies and practices and their intersection with indigenous people, who often inhabit the places targeted for conservation. "Nature," "environment," "biodiversity," and related terms have come to serve as banners around which to rally international environmental movements, which are grounded in laudable goals but often have the unintended effect of marginalizing local people or extinguishing traditional livelihood strategies. Many contemporary conservation projects stem from decidedly Western epistemologies—such as a vision of wilderness as a place unsullied by human activity—and are enacted by a combination of multilateral agencies, nongovernmental organizations, and state authorities. Extremely valuable recent work in this strand has explored conservation movements in Papua New Guinea (West 2006); the mobilization of science in the service of resource extraction and conservation in Indonesia (Lowe 2006; Tsing 2005; Peluso 1994); the linkages between conservation and community development in Mexico (Haenn 2005); and the political ecology of resource access in Madagascar (Gezon 2006). Much of this work shares the common goal of understanding what Arun Agrawal (2005) has termed "environmentality," the linkages between states and specific localities—mediated through power, knowledge, and discourse—and the resultant relationships between people and the natural world.

The other strand of environmental anthropology, what I call the "brown research agenda," examines the environmental and health risks associated with modern industrial life. Many development practices, particularly those associated with the industrial mode of production, result in deleterious environmental byproducts. These byproducts, in turn, are often disproportionately distributed to social groups with less political and economic power. A growing body of work seeks to apply anthropological theory and method to understand the nature of modern industrial hazards, the distribution of ill effects, and the ways that communities and other social groups organize themselves to mediate or mitigate the threats they face. These threats may be the catastrophic events of our time, like the meltdown at Chernobyl (Petryna 2002), the release of noxious gas at Bhopal's Union Carbide factory (Fortun 2001), or the slowly unfolding human tragedy of decades-long nuclear testing in the Pacific (Barker 2003). But they may just as likely be the more mundane, yet no less menacing, effects of everyday industrial activity on the residents of our cities and neighborhoods (Checker 2007, 2005).

It is not sufficient to simply understand the nature of the environmental risks produced by modernity. Indeed, the most interesting and productive studies within the brown agenda give us some insight into the array of strategies—often creative, sometimes successful, always culturally meaningful—that people use to address the problems that confront them. Oftentimes, environmental degradation opens up new spaces for disparate social groups to come together in an effort to represent themselves politically and to seek more just and equitable environmental outcomes (Pellow and Brulle 2005).

Because the brown research agenda deals with complex and sensitive issues that unfold over time, it is uniquely suited to the ethnographic research approach, which is characterized by careful qualitative research and longitudinal engagement. Yet environmental anthropology, perhaps owing to its roots in cultural ecology and ecosystem science, has been relatively slow to recognize the industrial mode of production as a suitable topic of inquiry. Meanwhile, the millions if not billions of people suffering from exposure to toxic industrial emissions receive scant attention, despite the quickening pace of globalization and industrial production around the world. I thus hope that this book can be part of a significant reframing of issues in environmental anthropology that places the brown agenda—which addresses how communities and individuals cope with the environmental hazards of modern industrial life—at the center of scholarly inquiry.

THE ORGANIZATION OF THE BOOK

Chapter 2 provides an overview of China's peculiar path to rural industrialization. I first examine the discourses and practices related to industrial development during the Maoist era and the subsequent years of economic reform, sketching in detail how Futian's factories grew to provide industrial inputs for Panzhihua Iron and Steel Company, a massive state-owned enterprise. Then I take a closer look at some of the particular development challenges faced by government cadres and local residents in Futian, including a historic legacy of economic marginalization and the problem of bringing development to an ethnically heterogeneous population.

In chapter 3, I examine the intricacies of factory management in Futian under China's Reform and Opening policies. Over the past decade, in accordance with central government policies that allow for greater private ownership of the means of production, Futian's factories were purchased by private investors, which has fundamentally changed their relationship to the township government and to local villagers. This process, coupled with the rise of China's "floating population," a migrant labor force that numbers more than 100 million people, has dramatically changed the employment structure and access to wage-labor opportunities for local villagers in Futian. I describe this restructuring process and some of its implications for township finances as cadres and villagers alike face an uncertain industrial future.

Chapter 4 deals with the environmental costs of China's path to industrialization. After briefly tracing the nation's growing energy demand and its high reliance on coal as an industrial, commercial, and residential fuel source, I turn to an examination of rural industry as a key contributor to the nation's environmental problems. As part of my fieldwork, I conducted scientific air-quality monitoring in Futian, where little systematic information about pollution levels previously existed. I present the results of these monitoring sessions and discuss the implications of observed pollution levels for community health, ecosystem integrity, and sustainability.

While industrial pollution is a material fact of life in Futian, it is also a sociopolitical problem that is constructed through public perceptions, discourses, and lived experiences. In chapter 5 I take up the question of environmental values by examining villagers' narratives about the effects of industrial pollution. Their concerns focused on the effects of pollution on their health and livelihoods but also tended to be linked with concerns about the dramatic changes taking place in Futian's political economy, including the privatization of factories, which resulted in the economic marginalization of many local residents. Environmental values and attitudes, I suggest, must be understood in light of villagers' different positions within this shifting political economy. I examine two quite divergent narratives of pollution and the environment. One represents agricultural and other non-industrial community members' recognition of the pollution problem and delineates the ways in which these people understand and cope with threats to their health, their livelihoods, and the integrity of the agro-ecosystem. The other narrative is representative of factory laborers, managers, and investors, who see things very differently. It becomes clear at the end of this inquiry that scholarly treatments of "environmental values" and "environmental consciousness" need to be framed in ways that are locally relevant and culturally salient.

Chapter 6 focuses on the processes and consequences of pollution enforcement in Futian. I illustrate how villagers, motivated primarily by concern for their livelihoods, contributed to a stricter enforcement regime that ultimately resulted in the closure of local factories by officials from the Environmental Protection Bureau. In tracing the process of pollution enforcement both inside the EPB and in Futian's industrial compound, I show how regulatory officials' decisions are guided by a range of factors, including central government policy, SEPA pollution-emissions standards, media coverage of Futian's environmental problems, and a civil-society movement grounded in "rightful resistance."

Chapter 7 describes the social and economic consequences of factory closures in Futian through a critical examination of the concept of sustainable development. While the central government in Beijing is increasingly adopting the international discourse of sustainability, agencies at the district and township levels have radically different ideas about the meaning of sustainability. These ideas, I suggest, are based on different models of sustainability with sometimes contradictory goals, such as preserving ecological integrity, promoting community development, and retaining vital industrial revenue.

In the concluding chapter I revisit the central themes of the book—environmental values, civil society, and the challenges of achieving sustainable development—and reflect upon their implications for thinking about some of post-reform China's difficult environmental problems. I also assess the future of China's trajectory of industrial development and explore some of the contradictions and tradeoffs it will entail.

2

THE DEVELOPMENT IMPERATIVE

I cannot sleep until we build the Panzhihua Iron and Steel Mill. . . . If capital is lacking, I will donate the royalties from my own writing.

—MAO ZEDONG

Development is the ultimate truth.

—DENG XIAOPING

THIS CHAPTER'S epigraphs, despite being attributed to figures from opposite poles of twentieth-century Chinese political history, are strikingly similar in the urgency they attribute to national development. Chairman Mao Zedong, the enigmatic leader of China's socialist revolution who ruled the country from the establishment of the People's Republic of China in 1949 until his death in 1976, viewed development in general, and industrialization in particular, as a matter of the highest priority. For Mao, economic development entailed overcoming what he perceived as the nation's backward, "feudal" (*fengjian*) past by leveraging science and technology in a march toward an imagined future. Mao's successor, Deng Xiaoping, was equally convinced that industrial development was the key to China's future competitiveness on the world stage, although the means through which he pursued this goal differed considerably from his predecessor.

For both of these leaders and for China's current leadership, development is a national goal with both materialist and normative aspects. To the degree that development policies aim to increase industrial output and improve living standards, these policies operate at a material level. At the same time, China's development policies are also laden with normative goals such as the creation of modern industrial citizens, the improvement of the "cultural quality [*wenhua suzhi*]" of the citizenry, and the realization of true economic competitiveness on the world stage. A common billboard, still visible in many Chinese cities today, features the smiling face of Deng Xiaoping above a bold caption: "Development is the ultimate truth [*Fazhan cai shi ying daoli*]."

China has one of the longest uninterrupted agricultural traditions in the world, with a history of wet-rice cultivation stretching back at least 5,000 years. While many of its neighbors were urbanizing throughout the twentieth century, China remained a primarily rural, agrarian nation. During the reform era, however, China transformed itself into an industrial powerhouse. Arguably the most dynamic part of this industrialization story was the rural industrial sector, which comprised millions of small-scale factories throughout the countryside. From the late 1970s through the late 1990s, rural China experienced the most dramatic employment shift in human history as more than 100 million peasants moved away from agriculture and into the factories.

Many social, political, and economic forces helped to shape China's industrial transformation. Although rural factories came to prominence during the reform era, important development strategies during the Maoist period provided the early impetus for rural industrialization. Futian's factories were established primarily to support the massive state-owned Panzhihua Iron and Steel Corporation. The linkages between local factories and the township government constituted a unique feature of the industrialization process and led cadres to depend on industrial revenue over the years as a "meal ticket" that allowed them to provide key goods and services to villagers, including graded and paved roads, government offices, and schools. These linkages between industry and local government, moreover, had the unintended consequence of contributing to pollution in the countryside, since cadres often zealously promoted the growth of industry at the expense of environmental standards. In Futian, villagers and cadres alike viewed industrial development and the revenues it generated as key to remedying the problems of ethnic, geographic, and economic marginality.

THE MAKING OF AN INDUSTRIAL CITY

The story of Futian's development is intimately tied to Panzhihua, a city most Chinese think of as an "immigrant town [*yimin cheng*]." The Great Proletarian Cultural Revolution (1966–1976), a spasmodic campaign designed to renew the nation's commitment to socialist ideals through the purging of intellectuals and other "bad elements," has been well studied for its displacement of millions of urban "educated youth [*zhiqing*]" to the countryside. Less well known, though no less significant for students of Chinese historical demography, is China's Third Front Movement (comprising two phases spanning roughly 1964 through 1971), during which Mao and other leaders, fearing the destruction of their nascent national industries in an imminent war with the Soviet Union, relocated key industrial assets from the east and northeast to the vast interior hinterland (Naughton 1988). Panzhihua, a mountainous region in southern Sichuan, became the site of one of the most storied Third Front industrial complexes.

The official history of Panzhihua, as reported in the *Panzhihua City Record* (Sichuan Province Panzhihua City Editorial Committee 1995), depicts the area before the Third Front Movement as barren, forbidding, and sparsely inhabited, with only seven Han households along this winding portion of the Jinsha River, the primary tributary to the Yangtze. This is a story often repeated by the early Han settlers themselves, who came to the area at the behest of the central government, making it difficult to tell where official historiography ends and local oral history begins. In any case, the official story of Panzhihua as an uninhabited wilderness seems designed partly to give the reader a mental picture of just how forbidding and remote the region was before the Third Front settlement and partly to provide a tangible illustration of the sense of desperation felt by the early settlers, many of whom left their homes in Manchuria to contribute to the effort, encountering a vastly different ecological and cultural landscape from the ones they had left behind.

Early government surveys of the area suggested that Panzhihua (then called "Dukou" after a ferry crossing on the Jinsha River) probably had a sizeable population before the Third Front movement, at least one-quarter of whom were ethnic minorities: Yi, Dai, Bai, and Miao who lived mostly in highland areas (Li 1995:88). Mr. Ma, a retired engineer at Panzhihua Iron and Steel Corporation, told me about the summer of 1965, when he and his wife moved from Chengdu, the provincial capital of Sichuan, more than 500 kilometers to the north, to become part of that first contingent:

I was part of the Young People's Corps [*qingnian dui*] back then. At first I couldn't bring my wife or family; they had to stay behind in Chengdu and wait until we had made some progress here. Of course, we were all poor back then, but the hardships we endured here were extraordinary. We didn't have anything, not even coal to cook with, and the hills were covered only in scrub brush that was hardly fit for burning. We used to put three stones together, light a fire in the middle, and use the heated stones for cooking. The hills were full of nettles, and we would all be cut and scraped after a day's work. We wandered around with just one set of clothes, a wide-brimmed hat to keep the sun off us, and a canteen. As for transportation, we had nothing but our own two feet.

These physical hardships were no impediment to Community Party ideology, however, which demanded the development of Panzhihua for reasons of national security. On March 4, 1965, Chairman Mao Zedong read a Ministry of Mining and Industry report recommending the construction of an iron and steel mill at Panzhihua; that very day he announced a plan to develop the region. His subsequent comments reveal the strategic importance with which he viewed Panzhihua: "I cannot sleep until we build the Panzhihua iron and steel mill. . . . If capital is lacking, I will donate the royalties from my own writing" (quoted in Jinside Panzhihua Bianweihui 1990:3). Mao's pronouncement was also based on early geological surveys in the 1930s, supported by more comprehensive survey work in the early 1950s, which confirmed that the landscape in and around Panzhihua was endowed with rich mineral deposits, including iron ore (both hematite and titanium-bearing titano-magnetite), coal, gypsum, and other minerals. Further exploration revealed that Panzhihua was also home to 93 percent of China's total titanium reserves as well as substantial deposits of vanadium, two important minerals for the production of military hardware (Shapiro 2001). A prominent statue displayed in downtown Panzhihua still bears the title *Fan Tai Zhi Ge* (Song of Titanium and Vanadium). Government publications refer to Panzhihua as a "treasure store ranking first under heaven" (Panzhihua Municipal Government 2003:8–9).

By any measure, the influx of new settlers to the area in the early years of the migration wave was astounding. The historian Judith Shapiro, who has reconstructed the migration history of Panzhihua, reports that in the first year alone, more than 41,000 migrants arrived. That same year, Panzhihua annexed several large people's communes from Yunnan and Sichuan provinces, redrawing the local political boundaries and putting more than 80,000 people under the administrative jurisdiction of Panzhihua. This brought Panzhihua's population to more than 120,000 in 1965.

Between 1965 and 1971, the city continued to receive a steady inflow of migrants, with an average net gain of 22,691 annually. After the initial dispatching of work units from nearby provinces, thousands of iron and steel workers and their families arrived from around the country, in particular from the great steel cities of Anshan, in Manchuria's Liaoning province, and Wuhan, in central China's Hubei province. This legacy is reflected in the diverse composition of Panzhihua's population; a taxi driver once jokingly told me that a person could learn the Manchurian dialect simply by spending time in this small corner of southwest China.

Workers built the iron and steel mill and soldiers from the People's Liberation Army constructed the Chengdu-Kunming railroad line, which ran through an outlying area east of Panzhihua called Jinjiang and which would transport the steel produced there to industrial and military factories. Summer temperatures reached forty degrees Celsius. The heat, combined with the physically demanding and treacherous work, contributed to an astounding 13 percent mortality rate among the newly arrived workers in the early years of settlement (Shapiro 2001). But dedication and ideology drove the workers toward their goal. Mr. Ma recalled the inspiring slogan of those early days: "Don't think of father, don't think of mother; until you produce iron, you can't go home [*Bu xiang die, bu xiang ma; bu chu tie, bu hui jia*]." In July 1970, construction on the iron and steel mill was completed and the factory began producing steel. The Chengdu-Kunming rail line was also completed that year. A warren of tunnels and twisting track through the mountainous terrain, the railroad connected this isolated region with the national planned economy.

Panzhihua in the twenty-first century, though by no means as provincial as it had been in the early days, is still a company town; tens of thousands of local residents are employed directly by the Panzhihua Iron and Steel Corporation (Panzhihua Gangtie, commonly abbreviated as Pangang). Pangang is located on the north bank of the Jinsha River and consists of a sprawling complex of industrial infrastructure, company-subsidized highrise apartments, shops, and restaurants; it is a self-contained city that many residents refer to as "steel village [*gang cun*]." When a shift ends or begins at Pangang, hundreds of employees clog the streets on their way either to the factory or back home, many wearing hard hats and denim jackets with the company logo—a bold "P" and "G" superimposed on one another, for Pangang—embroidered on the left breast pocket.

WARMTH AND FULLNESS

Although the primary impetus for the Third Front Movement was the strategic avoidance of military attack, Communist Party leaders had a secondary rationale: the creation of a national industrial infrastructure that was regionally self-reliant. It was for this purpose that the Panzhihua Iron and Steel Mill was constructed in close proximity to the raw materials necessary for production, including coal and iron ore. Futian Township was one of the areas annexed by Panzhihua in 1965 when the city was established. It is located at the far western edge of Renhe District, one of three administrative areas within the jurisdiction of Panzhihua Municipality (see the maps in chapter 1).[1] Because of its recent annexation from Yunnan Province, Futian is a bit of a geographical oddity. It sits on a narrow spur of land in Sichuan province surrounded on three sides by Yunnan. If one comes by bus or car from the Panzhihua city center, the road passes through several Yunnan towns, enters and exits Futian, continues over a mountain pass to the northwest, and reenters Yunnan within a few kilometers, a strange amalgamation of border crossings that is summed up in a local saying: "Breakfast in Yunnan, work in Sichuan, lunch in Yunnan."

For many centuries, villagers in Futian have supported themselves by exploiting a complex agro-ecosystem. A tenant-farming system typical of prerevolutionary China was well established in Futian by the early twentieth century, with poor minorities and Han Chinese both working as tenants for two large Han landlord families surnamed He and Lu (Harrell 2001:288). In the early years after the founding of the PRC, many villagers participated in violent "struggle sessions" against landlords and rich peasants, precipitating land reform, a key goal of Mao and other political leaders who sought to quell class antagonism by socializing the means of production. By the mid-1950s Futian, along with most of rural China, had undergone a full-scale communization of agriculture in which peasants were organized into a three-tiered system of state farming based on people's communes, production brigades, and production teams. The farming family as an economic unit ceased to exist, and those peasants who worked on communalized land grew crops in accordance with output targets set by higher administrative levels and earned work points that entitled them to a share in the harvest after state procurement requirements had been met. Township government documents depict the preliberation lives and livelihoods of local villagers, particularly the ethnic minorities, as downright Hobbesian in their harshness and brutality:

> Before liberation [1949], the productivity of the Yi people in Futian was low, the agricultural output was poor, animal husbandry was underdeveloped. There was no industry, no transportation infrastructure, no economic development. Education and cultural preservation were backward. Most people were illiterate, and medical care was very poor.
>
> (FUTIAN TOWNSHIP PEOPLE'S GOVERNMENT 2004)

The Dengist era brought yet another radical reshaping of social and economic life. The Household Responsibility System was implemented in Futian in 1982 and was administered by twenty-one small production cooperatives (*xiao zu*), which were the vestiges of the socialist production teams (*shengchan dui*). Cooperative leaders had the difficult task of allocating formerly collective land parcels, both irrigated paddy and dry fields, to peasant households. This was done largely on the basis of household physiology: families with more mouths to feed were given more *mu* of land, although the process could also be politically contentious at times since land parcels differed in soil quality and access to irrigation water.[2]

During the decollectivization process, farming households were given certificates that afforded them use rights but not full ownership rights over two types of land leased from the cooperatives: responsibility land (*zeren tian*) and contract land (*chengbao tian*). Responsibility land was allocated to households in exchange for an ongoing commitment to deliver a grain quota to the government, while contract land was leased through a bidding process to villagers who wished to expand their land holdings (Rozelle et al. 2005). The cooperative maintained the right to appropriate land within its jurisdiction when necessary. In keeping with the Marxist principle of collective ownership of the means of production, agricultural land could not be bought or sold by individuals; as a consequence, households that invested in infrastructural improvements such as irrigation had no guarantee that they would ultimately benefit from such investments in the long term. Futian's agricultural system is thus what might be thought of as a "smallholder" system, maintained by a combination of factors, including ecological and topographical conditions that make large-scale mechanized production difficult, central policy favoring collective land ownership, and a complex leasing system that makes consolidation practically impossible.

Over the centuries, Futian's landscape was profoundly shaped by human activity. Agricultural lands consisted of three types: irrigated paddy fields (*tian*), dry agricultural fields (*di*), and steeply graded lands used for animal forage that villagers referred to as "waste land." The rocky, mountainous terrain meant that rice paddies, which were located in the valley bottom near the stream, had to be maintained through an intricate terracing system. Terraces were constructed with walls of packed earth, and terrace edges were often sown with trees or crops such as sweet potatoes, broad beans, and peas, both to increase crop outputs and to prevent erosion. Dry fields, located higher up in the valley, lacked a permanent irrigation source and were intensively irrigated by farmers, a process that entails carrying water by hand from the stream or from irrigation canals. Land parcelization was a perennial problem; many villagers held a small plot of paddy and several *mu* of dry fields located hundreds of meters away, which meant that they spent much of their time traveling back and forth between land parcels. Waste lands, or "yellow mountains [*huangshan*]," which consisted of steeply graded land covered by grass and scrub, were unsuitable for agriculture. These mountainous areas were owned and managed by the township government, and a number of households that specialize in goat herding used them as forage for their herds.

China currently supports more than one-fifth of the world's population (approximately 1.3 billion people) with just 7 percent of the world's arable land. An average household in Futian cultivates only three *mu* of land (about one-half acre), including both irrigated paddy and dry fields. As a result, what little land is available is intensively managed and has been for many centuries. In addition to the summer rice crop, some farmers grow winter wheat in their paddies and corn on their dry fields. Villagers also grow dozens of plant and root crops (including bok choy, garlic, Chinese chives, peas, and sweet potatoes) and tree crops (including eucalyptus, hemp, mango, and pomegranate). Because of the township's latitude (approximately 27° N) and its mild climate, many species can be double-cropped or even cultivated year-round. By far the busiest time of year is between the spring and autumn equinoxes (*chunfen* and *qiufen*), as demarcated by the lunar-based agricultural calendar (*nongli*).

Every household keeps hauling and draft animals, including donkeys, mules, and water buffalo, whose dung is a highly valued organic fertilizer. Chickens and ducks are kept for eggs, and pigs provide the staple protein source for villagers. Several households specialize in carp aquaculture; the fish are raised in man-made ponds and sold regionally, while nutrient wastes from the fish ponds are used as a fertilizer. Poor village households burn crop residues such as corn husks as household fuel for heating and cooking.

For as long as any current villager can remember, nearly every household has used some variety of a multiple cropping system on its dry fields. This includes both intercropping (planting different crops together in alternating rows) and relay cropping (planting multiple crops in the same field in chronological succession). Li Jiejie's eldest brother, for example, often planted peas in autumn and corn in early winter on his land in Tangba village; by the time the pea vines were maturing, the corn stalks provided a climbing structure for the vines. In turn, peas provided excellent nitrogen fixation for the soil, reducing the demand for nitrogenous fertilizers. These are strategic adaptations that have evolved over many generations as a result of prolonged contact with the land, and they allow villagers to maximize the productivity of a relatively marginal agro-ecosystem.

As China's agricultural economy continued to liberalize under the Household Responsibility System, farmers in Futian began opting to cultivate less grain and other subsistence crops and focus instead on producing commodity crops such as mangoes, melons, and other fruits, which could be sold in local and regional markets (Tilt 2008). For the first time, some households in Futian relied solely on the market, planting their dry fields entirely with commodity crops for sale in the regional market and using cash profits to purchase food for household consumption. The shift to commodity crops was in many ways a rational response to the profit motive created by market liberalization: grain prices remained under the control of the central government, but nongrain commodity crops proved a lucrative option for farmers who were capable of gauging market demand and adjusting their cultivation practices accordingly.

Farmers tended to be grateful for the cash income and increased autonomy such policies provided them. In their discussions with me, villagers unequivocally viewed the Household Responsibility System in a positive light. One aging farmer told me:

It's much better than before. Ten times better. We barely solved the "warmth and fullness problem" [*wenbao wenti*] less than twenty years ago. Before that, we never had enough to eat. We had to grow rice, because that's what the government told us to grow. We couldn't eat other things, like corn. Also,

FIGURE 2.1 Villagers in Wuzitian transplanting rice seedlings in preparation for the growing season.

there was never enough meat to eat. One family of seven or eight people could only slaughter one pig per year, so we ate very little meat. Now, we grow what we want and we can just about eat as much meat as we want. We can grow things that bring in cash, too, like bananas and mangoes.

"*Wenbao*," which translates literally as "warm and full," is how many Chinese peasants describe the standard of living just above poverty, when basic needs such as food and shelter have been satisfied. This is a standard of living that most villagers in Futian have attained only recently; anyone over forty years old can relate tales of hardship in which their families survived on rice and little else. During especially hard times under the collective system, many peasant households ate forage grass along with their stock animals; one friend in Futian sardonically referred to the collective period as "the era of green shit [*lubian shidai*]."

While villagers generally tended to appreciate the political changes that had allowed them to improve their lot in life, they also tended to be keenly aware that their economic gains were relatively modest compared to other regions of China. Television programs depicting China's urban consumer lifestyle, complete with automobiles, cellular telephones, and socializing in coffeehouses, reminded people in Futian that national development was in many ways leaving them behind. As one Han farmer in Wuzitian village put it, "Agriculture is only good to a certain point. It solves the *wenbao* problem, but it doesn't make you well-off." In dismantling the system of social benefits that existed during the collective period, the liberal economic reforms of the past three decades have also fundamentally transformed the state-society relationship in rural China. Many villagers maintained that while their economic lives were far better than before, people had to rely on themselves now (*xuyao kao ziji*), since the government was no longer supporting them with work points or pensions. Reflecting on this fact, the farmer continued, "Things are better than before, but it's risky. You have to count on yourself. If you work for someone else, for a company or for the government, you can get a pension when you retire. But as a farmer, we don't have any pension. You have to rely on your own hard work."

The liberalization of China's agro-economy compelled farmers to participate in the market in order to provide a decent standard of living for their families. Investment in children's education was generally the largest cash expenditure for local households. Futian Township had four primary schools, which enjoyed heavy subsidies from the district and provincial governments, but families who wished to provide a middle-school and high-school education for their children had to send them to the district town of Renhe, where they paid room and board as well as tuition. This provided a powerful incentive for farming households to increase their cash income by growing commodity crops. One middle-aged woman from a farming family illustrated this point to me by relating the story of a local schoolboy, smart and industrious, who managed to graduate from high school in Renhe after considerable financial sacrifice by his family. He scored well enough on the college entrance exam to enter Panzhihua University, or perhaps even Sichuan University in Chengdu, but his family lacked the means to send him. After spending a year seeking government educational subsidies, which proved fruitless, the boy finally joined the People's Liberation Army in desperation, where he would at least be guaranteed three meals per day and a modest income.

THE NATIONALITIES PROBLEM

When Han settlers began arriving in the Panzhihua region in large numbers in the 1960s, they encountered a complex ethnic and linguistic mosaic with many local variants of the Yi language, as well as small contingents of the Bai, Miao, Dai, Zhuang, and other ethnic groups. Referring to this ethnic patchwork, many local villagers joke, "Go five kilometers in any direction and you can't understand the language," although this is not as true today as it was in the recent past. China considers itself a "unified, multiethnic state [*tongyi duo minzu guojia*]." In addition to the dominant Han majority, which constitutes about 92 percent of the nation's population, there are fifty-five "minority nationalities [*shaoshu minzu*]" that received formal recognition by the central government following an ethnic-identification project conducted between 1950 and 1956.[3]

The *shaoshu minzu* represent a special development problem for the government bureaucracy. On the one hand, their perceived "backwardness" provides normative justification for targeted development, economic assistance, educational subsidies, and national welfare policies. On the other hand, high concentrations of minority nationalities are perceived by many as a barrier to actually achieving true development. It was reiterated to me countless times by local minorities and Han alike that the southwest region remains poor because it is inhabited by ethnic minorities who are not only culturally inferior to the Han but who also lack the will and ambition necessary to achieve successful development. The *minzu* also raise the question of cultural and political autonomy within an authoritarian governmental system. Various places with high concentrations of *minzu* have received the designation of "autonomous region [*zizhi qu*]," "autonomous prefecture [*zizhi zhou*]," or "autonomous county [*zizhi xian*]," but the ability of officials within these entities to practice self-governance or to influence central policy remains extremely limited (see Rossabi 2004).[4]

Along with the neighboring provinces of Yunnan and Guizhou, Sichuan is one of the most ethnically heterogeneous regions of China, with more than a dozen of the nation's fifty-five recognized *minzu* (Chinese Statistical Bureau 2005b:44). The southwest region, which is among the most biologically and culturally diverse places on earth, is also one of China's poorest, long considered the ethnic, cultural, and economic frontier of the nation. In fact, none of the provinces in which *minzu* account for more than 10 percent of the population is counted among the ranks of high- or middle-income provinces (Wang and Hu 1999). In central government policy and discourse, this is commonly referred to as the "nationalities problem [*minzu wenti*]." Although it stems from cultural differences between Han and minority peoples, the nationalities problem is also economic and political in nature: How can the Chinese government achieve targeted economic development among the minority populations so that the western regions do not become a drag on the national economy and a source of social or political discord?

The central government has sought a number of solutions to the nationalities problem, most notable of which is the recent "Great Western Opening" policy (*xibu da kaifa*), outlined in several speeches by former president Jiang Zemin in 1999. The policy, which was touted as a central part of the tenth Five-Year Plan for economic development (2001–2005), has become an important, if somewhat disjointed, strategy for regional development. Focusing on twelve provinces and administrative regions, including Sichuan, the goal was to improve living standards in the west by focusing on the extraction of the region's abundant natural resources, including oil, natural gas, and valuable minerals (Chinese Statistical Bureau 2001:395).[5] It was hoped that this would be a step toward offsetting some of the vast disparities between the coastal region, with its Special Economic Zones and its access to global capital and markets, and the relatively impoverished interior.

The development model laid out in the Great Western Opening policy takes advantage of the unique assets of the western regions, including abundant natural resources and cultural heritage. Eco-tourism is part of the plan, as is a two-pronged emphasis on the development of agriculture and industry. The policy also explicitly addressed the nationalities problem, insofar as the *minzu* were seen to be linked to underdevelopment both because of the structurally embedded inequalities between them and the Han (access to education and fluency in standard Mandarin, for example) and because of the particular cultural practices and ethnic consciousness of the *minzu*, which impede acculturation and development (Lai 2002). The goal of Great Western Opening was to begin to close the gap between the west and the more advanced areas of China by emphasizing social and economic development in an environmentally responsible way. Part of the policy reads: "By the middle of the twenty-first century, the Western region will become a new and strong region marked by sound economic development, social progress, welfare, ethnic unity, and with beautiful and clean mountains and streams" (Chinese Statistical Bureau 2001:12).

Most of Futian's residents call themselves Shuitian zu (literally, "rice paddy people"), which distinguishes them from some of the highland minority groups in the area, particularly the Nuosu, whose rugged territory does not support rice farming. By official government accounting, however, they are part of the Yi *minzu*, an ethnic group of some 7 million people who traditionally spoke (and in some places still speak) multiple dialects of the Yi language, which belongs to the Tibeto-Burman subfamily of languages. While the Liangshan Yi dialect, spoken by Nuosu people in the Liangshan autonomous prefecture just north of Panzhihua, has an ideographic writing system, the Shuitian language has no orthography.

Documents published by the Futian Township government also mention the presence of small numbers of eight other *minzu*, including Dai, Bai, Man, Hui, Zhuang, Miao, Naxi, and Nisu. Prolonged contact and intermarriage with Han people over the last century has led to a dramatic decline in the use of the indigenous languages and a cultural assimilation that many villagers refer to as "Hanification" (*hanhuale*). Han and Shuitian villagers dress the same, apart from the colorful embroidered headscarves or slippers sometimes worn by Shuitian women; they eat the same foods; and they maintain ancestral tombs at the margins of their agricultural fields in the same way, by sweeping and decorating them at the *qingming* holiday.

In fact, many villagers could not understand why an anthropologist would possibly be interested in studying the Shuitian, since in their own minds they had become so completely acculturated as to be ethnologically uninteresting. Indeed, a wedding held in the predominantly Shuitian village of Tangba during the course of my fieldwork featured the usual Han customs of black-on-red calligraphy, gifting the bride and groom with money, and the bride's presentation of cigarettes to the groom's elder kinsmen. Nearly one hundred guests, young and old, sat at low tables drinking green tea, playing mahjong, and listening to Chinese pop music. The only vestige of "genuine" Shuitian custom in this affair was a circle dance, which was

FIGURE 2.2   A Shuitian bride and groom on their wedding day.

held into the wee hours of the morning on the final day of the three-day celebration and which featured about a dozen unique Shuitian dances.[6]

Nevertheless, it was typical for Han and minorities alike to attribute the backward state of Futian's economy to the low "cultural quality [*wenhua suzhi*]" of the Shuitian people. In her analysis of public discourse regarding migrant laborers in China, the anthropologist Ann Anagnost (Anagnost 2004; 1997) points out that many urban Chinese consider the low cultural quality of rural people in general as an impediment to the accomplishment of national goals and an affront to national pride. Cadres in Futian regarded the problem of industrial development as being doubly difficult, given the township's relatively poor, uneducated population of Shuitian people. Often, villagers themselves cited their own lack of education and skills (referring implicitly and sometimes explicitly to their lack of cultural quality) as the reason development targets remain out of reach. Indeed, many villagers over the age of forty were functionally illiterate, which I came to realize after several embarrassing attempts at asking people to fill out written survey questionnaires.

Mr. Tian Lizhong, a Shuitian cadre in the Office of Agricultural Technology and an accomplished amateur scholar of *minzu* culture, described Futian as hopelessly backward culturally, economically, and technologically. One night shortly before celebrating the advent of 2003, the year of the sheep on the Chinese zodiac, Mr. Tian invited my wife and me to a barbeque hosted by his family, where more than one hundred pounds of pig were roasted in the family's courtyard. Cuts of meat and fat were fried with oil and salt in a massive wok, while the organs, intestines, ears, and tail were slowly roasted over a charcoal grill. The blood was also mixed with oil and fried; after it had congealed, the guests consumed it with gusto.

To illustrate his point about the backwardness of the *minzu*, Mr. Tian told me a painfully comical story about a Yi woman who married into a Han family in the city, crossing with her nuptials both ethnic and geographical boundaries. She did not feel at all at home with her new family, especially with her mother-in-law, who, like the archetypal Han mother-in-law depicted in many famous Chinese literary works, forced her to do all kinds of household chores. With her father-in-law, things were even worse. The man was a midlevel cadre in an important government office and eminently concerned about his public image. To impress her father-in-law, the girl was careful to adopt some urban habits, even going to the trouble of applying makeup to her face daily before leaving the house. Her efforts met with severe scolding, however, when one day it was discovered that the lipstick she had been using was in fact the father-in-law's red ink pad, which he used to stamp official documents in his capacity as a civil servant.

In his role as a collector and purveyor of knowledge related to ethnic identity, Mr. Tian also recited to me a rhyming poem entitled "Old Brother Yi," which further illustrates how the Yi are perceived as naïve, juvenile, and unfamiliar with the trappings of modern, civilized life:

YI JIA GANGE

Yijia gange jincheng lai,
Mopi yishang shuirong xie;
Yangpi koudai li beiqi,
Kuqiao baba kai chulai.
Yijia gange jincheng lai,
Yangpi dabiao jiaopixie;
Cunkuan zheng'er suishendai,
Qiche manzai yuliang lai.

OLD BROTHER YI

Old brother Yi comes into town,
Plopping his burlap sandals down;
Shouldering his goatskin pack,
He pulls out buckwheat cakes for a snack.
Old brother Yi comes into town,
Clothes and shoes of goatskin sewn;
He closely guards his bank receipt,
And trades for a carload of surplus wheat.[7]

Ethnicity, of course, is a relational concept that is created and maintained through processes of negotiation. Some people in Futian resolutely referred to themselves as Shuitian, bristling with resentment at the official government classification of "Yi," while others embraced the broader Yi identity, given that it came with the entitlements afforded to an officially designated *minzu*. Li Jiejie routinely scolded her eight-year-old son, Xiaohua, for his disgusting table manners by calling him "Yi," which she equated to the Nuosu of Liangshan prefecture and other highland areas to the north of Panzhihua. The Nuosu were traditionally stratified into clan-based castes, some of which owned slaves up until the middle of the twentieth century. Li Jiejie referred to them as "barbarians" (*yeman*) who have a taste for raw meat and who eat rice by holding the bowl close to their faces and shoveling the grains in. On one occasion, when my wife returned to Futian after a trip to Liangshan to volunteer in a forest biomass-measurement project in several Yi villages, she reported that her backpack had been stolen along the route by bandits who had climbed atop the moving bus and cut the cords that held the luggage in place. Most of the Shuitian responded with a shrug: "That's what happens in places like Liangshan. The people are *minzu*. They're crafty, and they can't be trusted."

ENTER THE FACTORY, NOT THE CITY

Soon after the completion of Panzhihua Iron and Steel and the creation of Panzhihua Municipality as an urban center, another trend peculiar to contemporary China, and one with far-reaching economic and ecological effects, began: rural industrialization. At the beginning of the reform era in 1978, more than 90 percent of China's massive rural labor force worked in agriculture. By the 1990s, approximately one-third of the nation's 500 million rural laborers had put down their shovels and hoes and taken up factory jobs. Many local governments came to depend on industrial revenue to pay for teachers, pave roads, and provide other key community services.

Ironically, government leaders, including Deng Xiaoping himself, were reluctant to take credit for the growth of rural industry, preferring instead to view township and village factories as emblems of the people's own resourcefulness or as examples of development from below:

> In the rural reform our greatest success—and it is one we had by no means anticipated—has been the emergence of a large number of enterprises run by villages and townships. They were like a new force that just came into being spontaneously. . . . The Central Committee takes no credit for this. Their annual output value has been increasing by more than twenty percent a year for the last several years. Instead of flocking into the cities, the peasants have been building villages and townships of a new type. If the Central Committee made any contribution in this respect, it was only by laying down the correct policy of invigorating the domestic economy. The fact that this policy has had such a favorable result shows that we made a good decision.[8]

> (DENG 1994)

The truth is perhaps a bit more complicated. Although the rural industrial transformation is part and parcel of China's liberal Reform and Opening policies, its roots go much deeper into the Maoist period; in fact, no segment of the current market economy has escaped the effects of the planned economy that preceded it, although it is easy to overlook these historical precedents (Meisner 1996).[9] Rapid industrial growth began during the Great Leap Forward (1958–1961), a tumultuous period when China's economic planners funneled national resources and expertise away from agriculture and into industrial production. Chairman Mao insisted that every commune and brigade stop growing crops and instead build small-scale furnaces for steel smelting. His goal, driven more by socialist utopian idealism than by reason, was to outpace the steel production of Britain within fifteen years and thereby show the world the inherent superiority of both the Marxist development paradigm and the "spirit" (*jingshen*) of the Chinese people. Mao, viewing the nation's people as its ultimate resource, figured he could accomplish his goal by simply "lifting the lid, breaking down superstition, and letting the initiative and creativity of the laboring people explode" (quoted in Short 1999:480).

The results of the Great Leap were famously disastrous: rural areas produced steel of such poor quality that it proved useless for most military and industrial purposes, and an estimated 30 million excess deaths occurred around the country as fields lay fallow and grain production plummeted.[10] The seeds of rural industry, however, had been sown. Leftover from the Great Leap were the "five small industries," rural factories that produced key inputs for agriculture, including iron and steel, chemical fertilizers, farm machinery, cement plants, and electric power. These rural industries gradually became administratively distanced from the central government as demands were placed upon them to be self-sustaining in terms of tax revenue and production and in terms of providing employment opportunities and generating income for rural residents (Lin Qingsong and Byrd 1994).

After Mao's death and Deng Xiaoping's ascendance to power in the late 1970s, the nation was set on a path toward what top political leaders called "socialism with Chinese characteristics," a euphemism that in fact meant a retreat from socialist ideals and an espousal of market-driven capitalist development that would gradually wrest the economy from state control. In the early 1980s, as the Communist Party Central Committee launched the Household Responsibility System and granted increasing autonomy to peasants, the rural factories of the Maoist period became the de facto property of the townships and villages.

The labor picture in rural China had long been one of chronic underemployment; the nation's massive population, coupled with the socialist practice of guaranteed lifetime employment known colloquially as the "iron rice bowl [*tie wan fan*]," meant that the communes were full of redundant, unproductive workers. When agricultural production became the purview of economically rational smallholder farmers imbued with the profit motive, however, it became clear that unproductive labor had no place in a market economy. As a result, millions of rural workers were "freed up" (to borrow Karl Marx's phrase) with no particular place to go. Rural-urban migration has been the common response to such structural changes in many developing countries, but in China the central government bureaucracy had effectively been subsidizing urban workers for more than thirty years, providing state-guaranteed health care, education, and pensions for them and their families. To offer these services to a surfeit of new migrants would place an untenable strain on state resources. So government planners encouraged peasants to "leave the land but not the countryside [*litu bulixiang*]" and to "enter the factory but not the city [*jinchang bujincheng*]." In other words, peasants were pushed by a combination of policy and market directives to give up the agricultural livelihood while remaining in rural

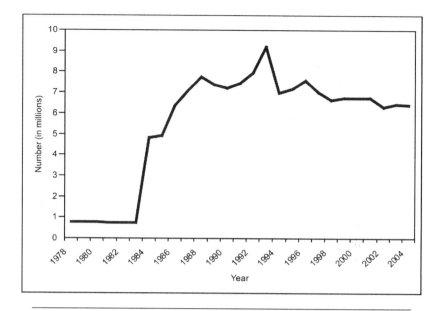

FIGURE 2.3 Increase in rural factories during the reform era, 1978–2004.
Source: Chinese Statistical Bureau 2005a.

areas. Thus China's rural industrial sector, which seems a contradiction in terms, was born.

Township and village governments responded swiftly by investing in nonagricultural enterprises. With tacit guarantees from the central government that it would leave rural industry largely unregulated, township and village factories grew tremendously during the reform era; they currently number 6.4 million and employ more than 80 million workers (see figure 2.3). These rural factories constituted the most dynamic part of China's economic reform story during the 1980s and 1990s, driving a profound transformation of the nation's employment structure.

A number of factors, both fiscal and cultural, help explain why rural factories have become such a vital part of China's industrial economy. First, government policy required rural factories to remit only a small portion of their profits as taxes to the central government, and this relatively light tax burden allowed them to put more revenue directly into industrial development (Naughton 1992). As a result, these small-scale factories were much more nimble and competitive than the lumbering state-owned enterprises that hailed from the period of high socialism. As rural factories remitted less tax revenue back to the central government, local governments at the township and village levels became more and more dependent on this money to fund all aspects of community development, including road construction, new schools, and salaries for government officials. As we will see in subsequent chapters, many township and village governments have been seduced by the lucrative nature of local factories, which serve as their "meal tickets [chi fan caizheng]." This causes all kinds of difficulties when market conditions change or when it comes to enforcing pollution standards.

Second, rural factories proliferated in part because local government officials in many cases benefited personally from industrial development. Government cadres must undergo mandatory yearly performance evaluations (kaohe zhibiao) that are based heavily on local industrial output. Because officials' careers depend upon the efficient promotion of local industrial development, they pursue the goal with great vigor. Ethnographic studies have provided some insight into this process. The anthropologist Mayfair Yang (1994), for example, in a study of the reciprocal social networks (guanxi) that are a central part of Chinese life, argued that these networks are a key element in the development of rural factories. Cadres are increasingly expected to use their social connections for procuring resources and obtaining the necessary land-use permits for local factories. According to Yang, the growth of rural industry during the reform period caused a displacement of the Maoist slogan of "serving the people [wei renmin fuwu]" in favor of a new slogan emblematic of the cadre's entrepreneurial role in local development: "serving the people's currency [wei renminbi fuwu]" (1994:161).

MOBILIZING FOR INDUSTRY IN FUTIAN

Futian's industrial ambitions had a modest beginning. Local factories initially played a peripheral role in the operations of Pangang, providing cheap inputs to the industrial giant. "Mayor Hu," a good-natured man who was actually the former mayor of the township, retained his civil-service title out of the continuing respect of the villagers. Approaching seventy years old, he lived on his retirement pension and farmed two mu of land "as a hobby" in Tangba, where his clan was one of three in the primarily Shutian village. In an interview with me, Mayor Hu described the township's industrial beginnings with a kind of paternal affection. Beginning in 1984, the township invested in two small coking plants and several coal mines. Coke is a hard, porous carbon material used for high-heat industrial operations, especially steel production, where its role is to burn off the iron content in the blast furnace. Workers extracted bituminous coal from the mines and transported it by truck to the coking plants, each of which housed several brick furnaces in which the coal was ignited and allowed to smolder for several days while the volatile compounds (including sulfur dioxide, tar, and oils) burned off. The crusted, silvery coke was then transported by dump truck on winding roads to Panzhihua, where it was sold on contract to Pangang. Mayor Hu and the other township cadres used the revenues from these two coking plants to expand their industrial base; by the time he retired from office in the early 1990s, the township boasted seventeen enterprises, including a gas station, a gypsum quarry, a construction company, several coal mines, a series of newly constructed zinc smelters, and even a driving school.

Villagers who chose to work in the factories were able to make several times the amount they could earn from farming. Most villagers welcomed the opportunity for wage labor since continued liberalization of China's economy meant that the central and provincial governments "retreated" from the provision of key services such as education and health care. Despite the cost of sending children to the city to be educated, families in Futian often saw this as a necessary investment, both as a way of fulfilling parental obligation and as a way of ensuring that the younger generation, bound by filial duty, would have the financial resources to care for them in their old age. In large part because of wage-earning jobs in local factories, the educational profile of Futian has changed dramatically over the past two decades: while it was uncommon for peasants over age forty to have more than a sixth-grade education, many young people have attended middle school and high school, and several successful families have even sent their children to universities in Panzhihua and Chengdu.

According to Mayor Hu, the communitarian ethic of the socialist period carried over into the early years of Reform and Opening. The 1990s were considered the "golden age [*jin huang shiqi*]" of rural industrial development throughout China, allowing industrial profits to be funneled into community-development programs. In Futian, new schools were built or renovated in all of the villages; a section of the winding, intervillage road was graded and paved; and the six-story government office building was constructed. Cadres invested more than 600,000 yuan in Tangba alone, the Shutian village that was the poorest among Futian's three villages. The benefits of industrialization, in short, accrued to the community as a whole, and even farming households without a direct financial connection to industry enjoyed a share in the fruits of development. This commitment on the part of government officials to community development was dubbed the "Futian Model" in policy documents from the region (Li 1995:237–66);

FIGURE 2.4. An elderly Shutian woman poses in front of her new satellite dish.

successful here, it could in theory be applied anywhere in the southwestern region, where the problems of poverty and ethnic backwardness were so often intertwined.

Futian's industrial development path proved quite lucrative. Nearly all the villagers recall the early years of industrialization with fondness. One of my favorite photographs taken in the township, both for its composition and for the story it tells, depicts a seventy-eight-year-old Shuitian woman in Wuzitian village, her grey hair tucked under a traditional headscarf called a *toupa* (see figure 2.4). She is standing in the courtyard of her house, squinting in the midmorning sunlight. To her right is a 200-*jin* (more than 200-pound) pig, which her husband and son have just slaughtered and hung upside-down from the rafters to drain the blood from a gash on its neck. Directly behind her is a brand new satellite dish, which she had installed in autumn of 2002. When I remarked upon the satellite dish, noting how strange it was to me that someone who lived in a three-room house with mud walls and few possessions should choose to spend her discretionary income in this way, she shrugged: "I like to watch Beijing opera."

Over the course of many more months of fieldwork, I came to view this woman's satellite dish and the dozens more just like it in the township, which seemed to sprout as quickly and prolifically as mushrooms after a rainstorm, as metaphors for the rapid and far-reaching changes that were taking place in the lives of the villagers. On the one hand, many elements of rural life seemed timeless and unchanging: the hours of tedious work in the fields; the parochial, inward-looking worldviews of the villagers; the rhythm of daily life closely tied to the turn of the seasons. Within the township's four villages, most residents lived in simple houses made of a foundation of hewn stones, mud walls, and a roof of ceramic tiles. Most lived in joint families, with multiple generations living either under one roof or in a cluster of adjacent houses. Women cooked relatively simple meals, using coal when the household budget allowed and burning biomass such as firewood or corn husks when coal supplies dwindled.

On the other hand, as the influx of satellite television illustrated, most villagers had experienced very concrete gains in their standards of living during the reform years. Some were able to lease residential land parcels from the township government in the newly designated open district in the center of town, where the main thoroughfare was lined with small shops, a few restaurants, a handicraft-furniture shop, a barber, and a sanitation clinic, which provided basic medical care and vaccinated children against hepatitis, typhoid, and other common communicable diseases. On these residential parcels, a few dozen families had built modern, two-story "dream houses," purchasing cement, corrugated roofing, and other construction materials and relying on family networks for labor. Such houses could cost as much as 20,000 yuan, not including the expenses required to decorate them and fill them with consumer goods such as television sets, refrigerators, and washing machines. A few enterprising households in the open district even owned "bread trucks [*mianbao che*]," tall vans with narrow wheel bases which resemble a baked loaf of bread and which were used to shuttle villagers short distances for a fee of a few yuan. Nutrition had improved for most residents, too: Li Jiejie, for example, was several inches taller than her eldest sister and a head taller than her elderly mother.

These gains, made possible by the Reform and Opening policies of the central government, had begun transforming the lives of villagers in this culturally and economically marginalized corner of southwest China. In this sense, the reform era represented the fulfillment of villagers' desires for development (*fazhan*), a concept that implied a gradual unfolding, a movement forward in time toward an eagerly anticipated modernity. Small-scale industrial development, which provided jobs and income for many local villagers, played a key role in this process. As one elderly farmer told me with obvious pride, "My son is not a farmer like me. He worked hard in several factories and now he's much richer than I am."

3

SAYING FAREWELL TO COMMUNAL CAPITAL

All Chinese people support the Communist Party and participate in the establishment and reform of the socialist system. This is because the Communist Party can bring them a higher level of riches and continually improve their material and cultural living standards. During World War II, people said, "Thirty mu of land, a cow, and warmth for my wife and children—this is Communism." After liberation, people said, "A two-story house, electric lights, and a telephone—this is Communism." After collectivization, people said, "Eating rice without paying for it— this is Communism."

—LIN LING AND COLLEAGUES, SICHUAN ECONOMISTS

China's biggest problem is the gap between rich and poor.

—FUTIAN FARMER

AFTER MAO Zedong's death and Deng Xiaoping's ascendance to power in the late 1970s, the People's Republic of China was set on a path toward what top political leaders called "socialism with Chinese characteristics [*zhongguo teside shehui zhuyi*]," a euphemism that in fact meant a retreat from socialist ideals and an espousal of market-driven capitalist development that would gradually wrest the economy from state control. "Socialism with Chinese characteristics" sounds like many other reform-era platitudes without real-world implications, but in fact it has had a very real effect on the structure of rural industry in Futian and throughout China. One of the hallmarks of economic liberalization in China today is the transfer of state-held assets into private hands, a trend that would have been unthinkable in Mao's time. But in the competitive marketplace of reform-era China, rural factories faced an array of economic pressures. Their role as collectively held assets meant that they served two purposes that were sometimes at odds with one another: generating profits, and providing employment security for township residents.

Accomplishing these two tasks proved difficult to sustain in the long term: after several difficult years, Futian's many factories began experiencing financial troubles that caused them to consolidate, privatize, or collapse, a process that was in line with trends in the rural industrial sector nationwide. In response, the central government has ironically facilitated the concentration of wealth and the privatization of collective assets by espousing policies that allowed for greater private ownership of industry, moving cautiously but steadily away from the Maoist ethic of the means of production being owned by the people (*quanmin suoyou zhi*) and toward a restructuring of industry called "grasp the large and let go of the small [*zhua da fang xiao*]." In practice, this meant nurturing the large and important state-owned enterprises while allowing smaller factories to fail or to privatize. By 2002, the industrial compound in Futian's open district had consolidated to the point where only three key factories remained: a zinc smelter, a coking plant, and a coal-washing plant. Each had been purchased by private investors, and this restructuring process fundamentally changed their relationship to the township government and to villagers. It also had important implications for township finances, employment opportunities, and migrant labor as the township faced an uncertain industrial future.

THE ZINC SMELTER

On a crisp but sunny day, Mr. Zhang, a soft-spoken man in his early sixties, sat on a low stool inside his two-room house, smoking Marlboro cigarettes. A color television set and DVD player rested on a table in one corner, with a stack of pirated Hollywood DVDs beside them. The house, constructed of red bricks and scavenged wood scraps, was located inside the industrial compound, and doubled as the base of operations for the zinc smelter. Mr. Zhang's wife used a corner of the main room for keeping the smelter's financial books; hand-written receipts were stacked in neat piles on a wooden desk, and several metal filing cabinets took up one wall. Mr. Zhang had migrated to Futian in 1998 from the neighboring province of Guizhou with his wife and their adult son. Together they had invested 450,000 yuan (roughly $55,000) in personal savings and bank loans to purchase the zinc

FIGURE 3.1. Workers shovel coal into the furnaces of Futian's zinc smelter.

smelter from the township government. The savings came primarily from Mr. Zhang's retirement account, which he had cashed out after serving as a high-school teacher for thirty years in the provincial capital of Guiyang.

Mr. Zhang's factory was considered a secondary zinc smelter (*jinxin-chang*). The local zinc-bearing ore, much of which comes from Huaping county across the provincial border in Yunnan, is a zinc-sulfide concentrate. Before 2001, the township government had owned and operated a primary zinc smelter (*chuxinchang*), which was used to remove the sulfur content from the zinc-sulfide ore, providing Mr. Zhang's secondary zinc smelter with a more pure form of ore for its smelting activities. The defunct, fifty-meter tall smokestack of the primary zinc smelter, which closed in 2001 when it became unprofitable and the township government could not find an investor to take over operations, is the most visible landmark of the in-dustrial compound. Mr. Zhang's secondary zinc smelter now acquired ore from Shilongba Township in Yunnan province.

The smelter consisted of six brick furnaces connected by a rudimentary ventilation system that drew coal smoke through a series of pipes and vented it through a smokestack. It typically operated six or seven days per week and produced thirty to forty tons of pure zinc per month, which were trucked several hundred kilometers south to the city of Kunming and sold to a company that manufactured galvanized roofing materials. The smelter consumed coal at an astonishing rate: approximately ten to fifteen tons of raw coal were burned in the furnaces for every ton of pure zinc produced. Mr. Zhang had initially come to Futian in hopes of improving the factory's profitability, selling it off within a few years and making enough money to secure his family's financial future. With enough time and effort, he told me, he hoped to be able to "pave a good road for my children [*po yi tiao lu wei haize zou*]." When operations were steady, his net income amounted to about 10,000 yuan per month, more than twenty times the township average. His eldest son, who also held a share in the zinc smelter, lived with him in Futian. His youngest son was getting advanced training as an engineer in the city of Guiyang, but the youngest son's wife, a thin, fragile-looking woman in her late twenties, had the task of staying in Futian with the family to prepare meals, do the washing, and perform the other duties of a filial daughter-in-law. The smelter employed twenty people, all of them men. Most were migrant laborers who either had social connections (*guanxi*) with Mr. Zhang back in Guizhou or who had prior experience working in similar industries nearby.

On this particular day, only four out of six furnaces were in operation. Wearing soiled clothing, their faces and hands blackened with soot, the men shoveled piles of raw coal into the furnace chutes. Other workers placed semirefined zinc ore in cylinder-shaped ceramic crucibles and hoisted them into the furnaces with metal tongs. After a period lasting from a few hours to more than a day, depending on the amount and quality of the ore, the men would carefully remove the cylinders and pour the molten zinc into rectangular molds. The hillside surrounding the smelter was littered with coal piles, empty ceramic cylinders, and unusable zinc slag. The men worked slowly and methodically, without masks or eye protection, despite the thick, sulfurous smoke pouring out from the sides and top of each furnace.

THE COKING PLANT

On a warm spring day in 2003, I rode alongside Mr. Tian, a coking-plant employee, to a state-owned chemical fertilizer factory in Huaping County, Yunnan. He steered the six-ton blue dump truck, laden with a twelve-ton load of coke, cautiously down the mountain roads into the wide, fertile valley, where farmers were beginning to level and clear their paddies with water buffalo in preparation for the rice season. We passed several groups of itinerant beekeepers, who transported their hives northward over the course of the spring, pollinating the flowering crops and fruit trees for a small fee from local farmers and selling their honey from makeshift road-side stands.

The fertilizer factory was a hulking tangle of pipes and smokestacks smelling of chemicals and sulfurous smoke. Several flare stacks, used for burning off waste gas, glowed brightly overhead. Mr. Tian steered the truck into a weighing station and received 3,000 yuan from an attendant, 250 yuan per ton of coke. According to the fertilizer factory's financial manager, a smartly dressed woman who bustled constantly from the main office to the site where coke was off-loaded, the factory produced 7,000 tons of chemical fertilizer during an average month, most of which was distributed to individual farmers and agricultural production teams in Sichuan.

On the winding road back to Futian, Mr. Tian told me that he had mi-grated to Futian from Yanbian County, in the northwestern part of Panzhi-hua Municipality, and had been working at the coking plant for more than two years. Driving trucks is an enviable occupation in the Sichuan country-side, where the back-breaking work of farming produces meager returns. Mr. Tian earned a good salary of 2,000 yuan per month. His wife and young daughter still lived in Yanbian, where he visited them once each month. When we arrived back at the coking plant, several dozen young men in soot-blackened clothes were unloading two dump trucks filled with raw coal. They hauled the coal in wheelbarrows and placed it in one of eleven ovens, each of which consisted of a brick-lined trench some eight feet wide, ten feet deep, and twenty feet long. Raw coal is converted to coke through a heating process in an oxygen-free environment. Futian's coking plant ac-complished this task in a low-tech fashion: the workers tightly pack each oven with coal, igniting it from beneath and allowing it to smolder for sev-eral days. After this, the men would pry away large chunks of porous, silver coke with crowbars, load it into the wheelbarrows, and fill the company dump truck for another run to the fertilizer factory. Four or five furnaces typically operated simultaneously while the others were either in a state of preparation or clean-up.

Until the late 1990s, most of the coke produced in Futian went to fuel steel production at Pangang. However, the giant iron and steel smelting facility has since increased its reliance on geothermal and hydroelectric power sources. The Ertan dam, currently the largest hydroelectric project in China pending the completion of the Three Gorges Dam, was completed in 1998 and has an installed capacity of 3,300 megawatts. It was constructed sixty kilometers outside of Panzhihua with a World Bank loan of US$1.2 billion

FIGURE 3.2. Smoke billowing from Futian's coking plant.

and is now the most important power source in the region. Consequently, Pangang's consumption of coke has plummeted, and industrial entrepreneurs in Futian have been forced to seek other markets for their product, chief among these being the state-owned fertilizer factory in Huaping.

The coking plant was established in 1984 as a collective enterprise (*jiti qiye*) with a total investment of 600,000 yuan, which the township government raised from villagers, tax revenues, and loans from the Agricultural Bank of China. Because Futian was classified as a minority township (*minzu xiang*), local cadres were also able to secure 1 million yuan in low-interest loans from the Panzhihua Minority Affairs Commission and the Panzhihua Bureau of Public Finance. Initially, the coking plant proved quite lucrative; township financial documents showed annual net profits of 50,000 to 100,000 yuan during the first few years of operation. By 1996, however, obtaining raw materials in Futian became difficult as the township's own coal reserves were depleted, and cadres decided to begin importing coal from Huaping County. Staggering beneath rising production and transportation costs, the coking plant closed down in 1997. By that time, it had accumulated a debt of more than 5 million yuan.

Township cadres saw privatization as the ideal solution to the coking plant's financial problems. Mr. Li, who hailed from the city of Chengdu and whom I saw only on rare occasions when he visited the township, purchased the plant in 1999. A middle-aged man with slicked-back hair and a cellular telephone clipped to his belt, Mr. Li entered into a convoluted financial relationship with the township government whereby he could hire his own workforce, consisting mostly of migrant laborers, and pay an output tax on the amount of coke produced, along with a land use fee, to the township government. These funds went to the township Office of Industrial Development to partially offset the township's industrial debt.

Concerned about the egregious air pollution emissions at the coking plant, the Panzhihua Environmental Protection Bureau contributed 250,000 yuan for environmental mitigation technologies, which Mr. Li used to build a tall smokestack designed to release factory emissions higher into the air and away from workers. In 2001, however, an explosion at the plant damaged the smokestack, rendering it inoperable. During all of the observations I made, coal smoke was released directly from small vents on the sides of each brick furnace; the smoke was so noxious that I had to use a handkerchief to cover my nose and mouth while conducting interviews at the plant.

THE COAL-WASHING PLANT

On a late spring day in 2003 Mr. Li, a wiry, middle-aged man who worked at Futian's coal-washing plant, sat in front of a panel of levers controlling the speed of a series of conveyor belts that sent a constant stream of coal through a small shack that housed the plant's machinery. Mr. Li was from Yanbian County in the northwest of Panzhihua and had been doing factory work for most of his life. The sun overhead was hot, and Mr. Li's station, covered by a roof and cooled by the constant flow of a watery coal slurry, was the envy of the other workers, all of whom labored outside under the glaring sun. Most of the workers wore wide-brimmed hats to protect them from the sun, but their skin was dark and leathery.

Until recently, most coal in Futian was extracted from one of several mines in the township. The coal reserves of China vary in quality and in "cleanliness," a term that refers to the percentage of pyretic sulfur contained in the mineral mixture. There are an estimated 80,000 small mines operated by county governments, township governments, or private owners throughout the country (Smil 2004:16). Coal from China's southern and southwest regions is notoriously high in sulfur content. The coal used in industrial boilers or domestic stoves is typically washed using a variety of

FIGURE 3.3. Catch basins filling with coal slurry at the coal-washing plant.

methods to remove some of the sulfur. Futian's coal-washing plant was not in operation throughout most of my field research period. On the first day I met Mr. Li and his coworkers, I had been climbing the hill to the industrial compound on my way to conduct an interview at the zinc smelter, when I heard a sound like the rumbling of a large truck. As I climbed higher on the hill, I realized that the sound I was hearing was the chugging of machinery at the coal-washing plant.

The plant began operations in the early 1990s as a collective enterprise. Similar to the coking plant, the coal-washing plant's profitability had declined in the late 1990s as the township's coal supply dwindled and more expensive coal had to be purchased from neighboring counties. It reopened during the spring of 2003 as a private enterprise whose owner lives in Yanbian County. The plant employed ten men, most of whom were also from Yanbian. The coal-washing process, which relied on heavy manual labor, was quite simple. Raw coal from Huaping County was shipped almost daily to the site by dump truck. The coal was particularly fine; I rarely saw a piece over one-half inch in diameter. The plant was situated on the natural slope of the hillside, and its chutes used gravity to pull the slurry of washed coal downward. Several workers stood at the top of the hill, using metal screens to sort the coal into piles according to diameter. Then they shoveled it onto a small conveyor belt, which ushered the coal into the machine shack, where Mr. Li sat at the controls. Water from a hose flushed the coal through rotating machinery, which agitated and mixed the coal into a slurry. The slurry traveled further down the hill through a series of narrow cement canals, where it collected in a catch basin about seven meters square, pooling to a depth of about one-half meter. The workers left the slurry to dry in the catch basis for at least ten days, removing it when it reached a pasty consistency. The washed coal was mostly used locally for household heating and cooking; villagers in the more developed open district of the township generally had indoor coal-fueled boilers that ran on small cylinders of coal called *meitan*, which had been formed from a mold. Other villagers farther away from the center of the township simply received the coal slurry in a bulk shipment, packing it by hand into little balls that could be thrown into small coal heaters.

THE FLOATING POPULATION

Just before taking a break for the Chinese New Year festivities in early 2003, township cadres met to discuss an unwanted byproduct of industrial development: the "floating population," migrant laborers who had moved to Futian to work in the factories. The meeting took place on the second floor of the government office building, in a large conference room with a cement floor and rows of wooden benches. Above the door hung a banner that read, "Develop family-planning policies; villagers control yourselves and implement a superior service for the community." Mr. Wang Chunyang, a short, anxious man with fastidious dressing habits and perennially shined shoes, was the Communist Party secretary of the township at the time, the highest-ranking local government official. At the beginning of the meeting, which Party Secretary Wang allowed me to attend as an observer, township cadres and approximately twenty members from Futian's four villager's committees (*cunweihui*) discussed a few budgetary issues. Then Party Secretary Wang rose to his feet and addressed the roomful of representatives.

According to Party Secretary Wang, the floating population raised two troublesome issues for the township. The first he referred to as the "cultural quality [*wenhua suzhi*]" of the local populace. "We cannot allow outsiders to come to Futian and lower the cultural quality of our citizens," he said, referring to the common practice of migrants squatting in the community, producing too many children, and, lacking the proper residence permits for government-subsidized education, keeping their children out of local schools. Waving his hands for emphasis, he continued, "We cannot afford to lower our standards."

The second issue was strictly fiscal. If township cadres were seen to be lax on family-planning policies, they ran the risk of losing future subsidies from the district and provincial governments. According to Party Secretary Wang, this would set the township behind on its development targets. Most of the other township cadres and village representatives were in agreement, immediate consensus being a common feature of government meetings in China. The meeting concluded with a resolution that the migrant families would be dealt with in short order.

China's central government has developed two incredibly efficient population-control measures. One, the "planned-births program," or *jihua shengyu*, was designed to control population quantity. The planned-births program, which is often erroneously referred to in the West as the "one-child policy," strictly limits urban households to one child but provides less-stringent regulations for rural areas with large minority populations and areas where a significant percentage of residents live below the poverty level. (These two conditions, in fact, often coincide.) Families in Futian, most of whom relied on labor-intensive agriculture to make their living, were allowed two children; however, they were still forced to comply with strict birth-spacing laws that required a couple to wait at least five years from the birth of their last child before having their next. The central government's goal is to gradually drive down the long-term fertility of rural women without entirely stripping families of their labor sources and inciting political turmoil.

Family-planning policies and population-growth targets are set by officials in the central government but implemented and enforced at the township and village levels. In Futian, a woman whom villagers referred to as "Dr. Zhang" was the staff member in charge of the family-planning clinic, which was located on the first floor of the building where my wife and I rented several rooms from Li Jiejie. In fact, it wasn't until several weeks after settling into our living space that I bothered to read the sign above one of our rooms, which read "Recovery Room No. 2 [*kangfu er shi*]." Dr. Zhang, who in fact had only a high-school education, oversaw the planned-births program by educating villagers on birth control options, distributing condoms and oral contraceptives, and assisting a visiting healthcare professional who conducted medical procedures such as installing intrauterine devices or performing tubal ligations and vasectomies. In extreme cases, Dr. Zhang helped perform permanent sterilizations on unwilling villagers at the behest of the township government, although these more draconian measures were reserved for those who flagrantly violated policy and refused to curb their own fertility.

The financial penalties for violating family-planning policies were severe: bearing a third child in Futian triggered a 21,000 yuan fine, an amount equal to four years' income for an average family. Failure to meet the child-spacing requirements also resulted in a fine, although Dr. Zhang could not be specific about the amount. Overall, the family-planning policies have been highly successful: although China's population currently numbers more than 1.3 billion, the natural rate of population growth has fallen from 2.5 percent in the 1970s to less than 1 percent currently. China's total fertility rate, the average number of births per woman, fell from nearly five to under two during the same period (UNDP 2007).[1] The central government's goal is to hold population growth down to 1.4 billion by the time of the 2010 census, and most estimates project that it will grow to 1.6 billion before reaching a peak in the middle of the twenty-first century. Fertility rates in Futian have also fallen precipitously in recent years. Li Jiejie told me on several occasions that nearly everyone had large families before the family-planning policies were instituted in the late 1970s, in part because child-hood mortality was so high; three of Li Jiejie's eight siblings had died before reaching the age of fifteen, two of them from food- and water-borne pathogens such as hepatitis and cholera.

The other population-control mechanism, the household-registration system, or *hukou*, was designed to limit the mobility of Chinese citizens during the socialist period. Legalized in 1958, the *hukou* system required every citizen to register at birth as either an urban or rural resident and to declare a fixed place of residence. Until quite recently, rural and urban residents were subjected to wholly different state controls and were entitled to different state benefits. Urban household registration was closely linked with an "urban public goods regime" (Solinger 1995), which included subsidized housing and food rations, employment opportunities, health care, child care, and retirement pensions. The welfare of peasants, meanwhile, was the responsibility of the rural collective. To contain the strain on state resources brought on by this system of uneven entitlement, rural-urban migration was severely restricted.

Meanwhile, how was the Chinese political leadership to promote the forward push of industrial development when the mobility of its labor force was so severely curtailed? Township and village enterprises offered the perfect solution in the form of industrialization without urbanization. However, the influx of global capital into China, most notably to the dozen or so

Special Economic Zones on the east coast, presented the command economy with another problem. Foreign-owned factories, which helped to keep China's GDP growth rate humming along at nearly 10 percent per year, demanded plenty of cheap labor, which has proven China's comparative advantage in the global economy. Policy accommodations were made once more. The *hukou* system was gradually relaxed both through formal channels, including allowing temporary residence in urban areas for those holding a special "blue-stamp *hukou*," and through informal channels, which entailed lax enforcement of existing policies. The result was the creation of China's "floating population [*liudong renkou*]," an ethereal migrant workforce estimated to number more than 100 million (Zhang 2001; Solinger 1997).

This estimate almost seems conservative when one sees the flood of laborers on the streets of provincial cities like Chengdu and Kunming: midnight garbage collectors who sift through every trash can in search of recyclable items; shoe shiners who charge two yuan for their services; telephone-card hawkers; food-stall operators; construction workers; train-station porters; female greeters at upscale restaurants wearing mandarin *qipao* dresses. Lacking formal education and skills, the floating population is the perfect embodiment of the proletariat, an underclass with nothing but its labor to sell (Sun 2000). Migrant workers are also the subject of a rich discourse of antagonism toward those in Chinese society who are rootless; they are seen as unwashed, ungovernable, and, as Party Secretary Wang's comments illustrated, of low cultural quality. Because they are not rooted in place, they are also seen as deviants from the ideal family type in traditional China, which is a joint family consisting of four generations under one roof (*si shi tong tang*).

As cadres in Futian came to realize, "floating" was not merely an urban phenomenon. Although China's cities present migrant workers with the strongest attractions—the proposition of cleaner work, higher wages, a taste of modernity, and the chance to remit money back to their families—towns of almost any significance have their migrants. The prospect of finding jobs in rural factories drives much of this intra-rural migration, and Futian itself was home to some hundred migrant workers, many of whom came from Guizhou, another underdeveloped mountainous province in the southwest, bringing their families with them. They lived in rows of squatter houses made of scavenged and improvised building materials inside the industrial compound on the steep hillside on the edge of the open district. These migrants presented a peculiar problem for the township government: as products of the loosening *hukou* policies, they were by law still residents of their respective hometowns, which made it difficult for local cadres to tax them or to enforce family-planning policies.

The Yang family, whom I met while interviewing other workers in the zinc smelter, are a case in point. Mr. Yang, who lived with his wife and three children in a dilapidated house constructed of wood scraps, had found employment in the smelter through his social connections (*guanxi*) since he hailed from the same province as Mr. Zhang, the smelter's primary investor. The trees surrounding their house, mostly eucalyptus and Yunnan pine planted within the last twenty years as part of the national "Grain to Green" reforestation campaign, were dying from the sulfurous smoke of the factories. Inside the small house, on a floor of packed earth, was a single bed, its wooden frame made by hand, with a mosquito net hung from overhead and tucked neatly beneath the mattress to protect the family from mosquitoes and other endemic pests, such as funnel-web spiders. There was also a black and white television set and three or four low wooden stools. A small coal-fired cooking stove sat in one corner.

Over a simple dinner of chicken broth, stir-fried cabbage, and rice, Mr. Yang, through a chain-smoker's yellow teeth and a thick Guizhou accent, described the hardships of his previous life in his home village outside the provincial capital of Guiyang, which had led him to seek factory work. Like most migrants, he described his decision to "go out and work [*qu waimian dagong*]" as the product of a demanding agrarian livelihood that provided no opportunities for improving his lot in life. Though its hours were long and its working conditions sometimes inhuman, the factory offered Mr. Yang what he most desired: a reasonable income and the prospect that his children's lives would be better than his own. He had few complaints about life in Futian, apart from the harsh treatment he received from township cadres eager to enforce the family-planning policies following the government meeting: "Party Secretary Wang wants us to pay a fine. He claims we're a burden on the local government." His wife, a plain-looking woman in her late twenties, continued in the same Guizhou accent as her husband, "They've been hassling us for some time now. They say that we have to pay 2,000 yuan because we've broken the family-planning policies." The sum, although far less than the maximum amount of 21,000 yuan that cadres could have demanded, still amounted to about four months' income for the family.

Two of the family's three children, a six-year-old boy and a four-year-old girl, sat on the stoop of the house, eating their soup and picking awkwardly at their rice with chopsticks. The youngest child, an eighteen-month-old boy who was evidently the source of the controversy, was sleeping soundly beneath the mosquito net on the narrow bed. The couple jokingly referred to him as *guizi*, "the expensive one." The source of the conflict was that the family had resided in Futian for more than two years, so it was clear that the boy had been born in the township. Cadres claimed that the couple should rightfully be subject to the standard fine but, acknowledging the family's grim financial situation, agreed to accept a lesser amount. For their part, the couple admitted that the baby had been born in Futian, but claimed that because their *hukou* was still legally in Guizhou they were effectively outside the jurisdiction of the local government. This argument had a certain logic to it, given that the family received no social services—no elementary education, no health care—from the Futian government. I asked what would happen if the family refused to pay the fine. "Refuse?" said Mr. Yang's wife. "We're not refusing to pay it; we can't pay it."

The family finished eating and Mr. Yang's wife began washing the bowls in a plastic basin outside the front door. The air was beginning to chill as the sun set behind the mountains. The smell of sulfur from the factories was still overpowering. On my way out the door, Mr. Yang said under his breath, "Two thousand yuan." He shook his head and laughed, "Where would we get that kind of money?"

THE FACTORY AS MEAL TICKET

In my experience, there are at least two kinds of cadres in rural China: ideologues, who toe the party line and tout its latest slogans, eager to show their best public face to the world, and pragmatists, who place the accomplishment of concrete tasks above all else and who pursue their occupation with a measure of critical reflexivity and even a bit of levity. Little Hu was the latter. A young Shutian man with a round face and a flat-top haircut, he ran the township Office of Industrial Development. He was the son of "Mayor Hu," whose family hailed from Tangba, the predominantly Shuitian village in the southwest of the township, where his clan was one of three in the village (the others are the Ni and the Li).

Sitting behind his cluttered desk on the second floor of the government office building, watching the plumes of black smoke rise from the factory compound on the hillside, Little Hu told me a joke. "Bill Clinton, Boris Yeltsin, and Deng Xiaoping are driving down the road," he said, making reference to the 1990s. "They're on their way to some important banquet or state function. Of course, each leader is riding in his own chauffeured, black limousine. The cars approach a fork in the road. Bill Clinton's car signals to the right, slows down, and makes a right turn. Boris Yeltsin's car also signals to the right and makes a right turn. Then Comrade Deng's car"—Little Hu began to grin, giving away the punch line—"signals left, then makes a sharp right turn."

That was the first time I had heard that particular joke, but it was not the last. In fact, rural Chinese often spoke of "socialism with Chinese characteristics" with a glint in their eyes and a slight smile on their lips, aware of the irony of the phrase. In reality, most sectors of the Chinese economy are controlled by market forces, but the market is in turn guided by the heavy hand of the party-state, at both the national and local levels. Ironically, the Chinese Communist Party's legitimacy rests squarely upon its ability to effectively use the discourses and economic tools of the capitalist market.[2] Many government cadres told me openly that there was essentially no difference between socialism with Chinese characteristics and capitalism but that even though much had changed in contemporary China, one could not speak of China as capitalist without encountering serious political repercussions.

In a more serious tone, Little Hu told me that the local investment climate had taken a downward turn over the past several years. In the late 1980s and early 1990s, when Little Hu's father was serving as mayor, profits and tax revenue amounting to more than 2 million yuan poured in each year from the factories, all of which were still collectively held by the township government. Cadres referred to local factories as their "meal ticket [chifan caizheng]," and used the revenue to construct the six-story government office building in the heart of the open district, pave the main throughway from Panzhihua, build two new schools for the poorest and most remote of Futian's four villages, and invest in new agricultural technology. Cadres frequently boasted to me about the government's success in providing the villagers with the "three connections [san tong]" essential to modern rural life: roads, water, and electrical power. By the end of the 1990s, every villager in Futian could truck their crops to market on reasonably well maintained roads, drink well water or water pumped from a government-subsidized groundwater system, and chat with friends and family into the evening beneath electric light bulbs. These constituted the tangible gains in living standards made possible by industrial development.

Structural changes within China's rural industrial sector altered the relationship between local government and industry in Futian. During the golden age of the 1990s, revenue from Futian's factories constituted about 85 percent of the township's operating budget, and cadres had relative autonomy to allocate these funds to community-development projects as they saw fit. By the late 1990s, however, Futian's factories encountered many of the obstacles endemic to the rural industrial sector throughout China. These included wasteful, inefficient production practices, and sometimes overt corruption; a lack of sufficient capital to expand industrial capacity; and problems securing loans from the state-run banking system, which ironically began favoring privately owned firms in its lending practices (Oi 2005). In Futian, these macroscale problems were compounded by several local phenomena. Coal stocks in the township had become depleted; a series of abandoned mine shafts scarred the hillside to the north of the open district. Factory managers were forced to truck in coal from across the provincial border, which increased their production costs. Furthermore, township cadres had made a series of bad investment decisions and faced mounting debt. In the eyes of local cadres, factory privatization, which seemed an ideological contradiction in an ostensibly socialist system, was seen as the only way out of their rapidly compounding financial difficulties.

But such radical changes in the political economy did not happen in a social vacuum; rather, the ways in which assets changed hands were shaped by the unique cultural and political landscape. The decade of the 1990s was like a fire sale. In many cases, local governments sold communal factories to insiders with connections to local government or industry or both, a phenomenon called "insider privatization" (Li and Rozelle 2003). In one recent case study I worked on in Yunnan province, for example, my colleague, a Western-trained Chinese anthropologist, observed as a state-owned sugar-processing mill was bought by a group of investors that included the mill's former manager, three deputy managers, the Communist Party secretary of the county, and the chairman of the state-sponsored labor union (Li and Tilt 2007). From a pragmatic standpoint, this makes good sense, since insiders have a working knowledge of production activities and possess the necessary social connections (*guanxi*) that are indispensable to greasing the wheels of China's complex bureaucracy. But such practices also contribute to poor management decisions, inefficiency, and outright graft and corruption.

I noticed, however, that the privatization process in Futian did not follow this pattern. Despite more than a decade of industrial growth, villagers remained comparatively poor. No individual or group had sufficient capital to purchase the township's industrial assets, so cadres began courting outside investors. The district government in Renhe published a series of glossy brochures to advertise the advantages of doing business in Futian, including rich mineral deposits and an abundance of labor. In addition, potential investors in local factories were offered a one-year tax holiday and reduced taxes for up to three years.

Futian's three factories—the zinc smelter, the coking plant, and the coal-washing plant—had all been purchased by outside investors by 2002. These investors now controlled the direction of local industrial development and profited from industrial activity, a process that the geographer Joshua Muldavin has described as the "dismantling of communal capital" (Muldavin 1996). Factory privatization did not mean that the township would no longer benefit from industrial revenue, but it did cause township cadres to radically restructure the relationship between the government and local factories. In a series of convoluted financial documents, the township government outlined how infrastructural assets (including factory equipment) would be purchased by investors, who would be responsible for paying a set of fees to the government. Mr. Zhang's zinc smelter, for example, was assessed a monthly management fee (*guanli fei*) of 400 yuan per furnace, a monthly tax (*shui*) of 150 yuan per furnace, and an annual land-use fee (*tudi fei*) of 2,000 yuan for each of its six furnaces.[3] The coking plant and coal-washing plant also paid similar taxes and fees to the township government. Despite this industrial restructuring, the financial outlook of the township was still bleak. While factory taxes came in on a monthly basis and provided sufficient operating revenue, the township still owed millions of yuan to various creditors, including Renhe District and the Panzhihua Bureau of Public Finance. In their negotiations with the private investors, cadres had failed to get the investors to agree to assume this long-term debt.

Most significantly, township cadres allowed the private investors to hire their own labor. In most cases, investors brought in workers from their home provinces or members of their extended kinship networks; in some cases, they simply culled from the ubiquitous floating population. The dismantling of communal capital was thus compounded by the outsourcing of labor opportunities. In the process, many villagers in Futian were laid off, although the township's registered unemployment rate (*dengji shiye lu*) remained low because cadres listed the laid-off workers as temporarily furloughed (*xiagang*), a common practice that made the township employment records look better during inspections by district officials.

In late 2002, the township government held its semiannual People's Congress (*renmin daibiao dahui*), in which representatives from all four villages were present to discuss the year's progress and outline future development goals. A white banner at the back of the conference room featured Deng Xiaoping's famous saying, "Development is the ultimate truth [*fazhan cai shi ying daoli*]." Another banner, which Party Secretary Wang had composed himself and printed out on one of the government computers, featured an equally optimistic, if more convoluted message: "See clearly the true situation and the new responsibility to take steps to beckon commerce, attract investment, and push Futian toward new development [*renqing xincai lieming juexin renwu nachu xinjucuo yi chao shang qi zi tuidong futian xin fazhan*]."

Party Secretary Wang had invited me to attend the People's Congress, so I sat on a bench at the back of the conference room and tried to be as unobtrusive as possible. Party Secretary Wang called role to determine that a sufficient number of representatives were present from each village. A total of forty-four village representatives were in attendance, all but three of them men. Most representatives were dressed in their dusty farm clothes and appeared as if they had just come in from the day's hard labor. Many of the men were smoking Red River cigarettes and reading newspapers or distractedly picking the dirt from beneath their fingernails. A tape recording of the national anthem was played over a loudspeaker, and the crowd rose to their feet to gaze at the flag at the front of the room. Following the national anthem, Mayor Zhang read the development progress report (*fazhan jinbu baogao*), which consisted of a comparison of key development indicators (including agricultural production, industrial output, and household income) between the 2001 and 2002 fiscal years. Not surprisingly, given the financial difficulties with the factories, the report reflected a general downward trend. Li Hong, the vice mayor, then proceeded to read the development targets (*fazhan mubiao*) for 2003, which cadres had devised over the previous months. The targets all represented slight increases over the previous years.

The heart of the meeting was an hour-long discussion of Futian's recent failure to meet satisfactory development goals, particularly in terms of industrial output. Party Secretary Wang facilitated this discussion, which included a brainstorming session in which the village representatives were encouraged to write down suggestions about how to continue economic development in the face of declining factory revenues. At the direction of Party Secretary Wang, a cadre distributed sheets of paper to each representative, and the room fell silent as people thought about their suggestions. I noticed that most people wrote a few lines, but when it came time for representatives to share their ideas with the group, a hush fell on the room. No one wanted to speak up. The whole point of the brainstorming session, prodded Party Secretary Wang, was to get input from the common people, or the "old hundred names [*laobaixing*]." Frustrated at the lack of response, which seemed to me the product of fifty years of tumultuous rural politics during which outspoken villagers risked serious retribution, Party Secretary Wang launched into an impromptu speech in which he implored village representatives to find a way to continue developing the township:

> Remember, we are in danger of falling behind our neighbors, Shilongba and Daxing [two townships nearby in Yunnan]. This shows the need to work harder to attain a *xiaokang* [well-off] society. We must overcome our backwardness. . . . The industrial output of Sichuan as a whole is immense. You think it's not possible for us to attain this level of development? You think it's too hard, that we're too poor, too backward? Well, I say that we old hundred names have knowledge of our own. We have experience. If we think we can accomplish something, then we can. All you have to do is liberate your thinking [*jiefang nimende sixiang*]!

In the hopes of helping cadres to liberate their thinking, Party Secretary Wang reported that he had ordered several dozen subscriptions to the *Panzhihua Wanbao* (Panzhihua Evening News), which he hoped would keep cadres abreast of current events and economic trends. His speech struck me as an interesting ideological collage. "Liberate your thinking" had been Deng Xiaoping's exhortation to the Chinese populace, by which he meant that China's old thought patterns, including both the feudal past and the period of high socialism, were impediments to liberal economic development that had to be abandoned if the nation was to achieve real gains. The concept of *xiaokang*, which literally means "small comfort," first appeared in the "Record of Rites" (*Liji*) from the Warring States Period (475–221 B.C.). This historical text describes an idyllic age of "great equality" (*datong*) in which heaven and man were in harmony and wealth was distributed evenly. But *datong*, according to the text, ultimately proved to be an unattainable goal because of the tendency of humankind to acquire material wealth for personal gain. The alternative to equality was said to be *xiaokang*, a state in which individuals sought their own pecuniary benefit. Lest the historical context of *xiaokang* seem ambiguous and a bit esoteric, Deng Xiaoping proclaimed in the 1980s that the *xiaokang* lifestyle required a minimum per capita income of 800 U.S. dollars, creating a definition of development that was at once tied to the shared ideals of the national history and grounded in a decidedly modern economic context.[4]

The role of capitalist markets and the restructuring of the rural economy had figured prominently in the Dengist regime's plans for actualizing development. After all, the economic reforms were predicated on the idea that "some must get rich first" in order that others may follow. More recently, however, top political leaders including former president Jiang Zemin and current president Hu Jintao have invoked the concept of *xiaokang* to acknowledge one of the chief acts of civic violence resulting from Reform and Opening: the creation of huge wealth disparities within the nation's populace. Income inequality, both between rural and urban areas and within communities, is one of the greatest threats to long-term social and political stability (Riskin, Zhao, and Li 2001; Khan and Riskin 1998). Current leaders have thus set a goal of attaining what they have termed "all-around well-being [*quan mian jianshe xiaokang shehui*]" for all of the nation's citizens.

Later in the People's Congress, Mayor Zhang, a local Shuitian man in his thirties, also underscored the need for continued development in the face of economic troubles, although the tone of his speech was more pragmatic than Party Secretary Wang's and relied less on the tropes of Dengist ideology:

> Look at the overall development situation of China. We have one of the largest economies in the world. And you have to say that's not bad. But we need more development of the economy, more development of technology, more development of culture. Look at the economy of China's eastern regions such as Guangdong and Shanghai. You have to say that there's a big difference between them and us in terms of development. We've only just solved the warmth and fullness problem [*wenbao wenti*]. We have a long way to go, and we have to depend on ourselves to solve our economic problems. We can't rely on the national government, or even the district government. It is only when our living conditions here in Futian are improved that we can begin to solve the really important problems, like education and cultural quality.

Cadres such as Party Secretary Wang and Mayor Zhang had faced a dizzying array of social and economic changes in a relatively short time period. As part of the liberalization of the local economy, they had presided over the privatization of the zinc smelter, coking plant, and coal-washing plant. Navigating these changes meant facing a number of new challenges that would have been unthinkable during the Maoist period. Cadres had to enter into complex financial relationships with private investors, regulate the flow of migrant laborers, and learn to cope with a new kind of morality that prized individual benefit over the collective good—all in the name of "socialism with Chinese characteristics." For Party Secretary Wang, Mayor Zhang, and their colleagues, more challenges were to come, including dangerous levels of air and water pollution from local factories.

4

THE ENVIRONMENTAL COSTS OF PROGRESS

China's environmental crisis has arisen, basically, because our mode of economic modernization has been copied from Western, developed nations. In twenty years, China has achieved economic results that took a century to attain in the West. But we have also concentrated a century's worth of environmental issues into those twenty years. While becoming the world leader in GDP growth and foreign investment, we have also become the world's number one consumer of coal, oil and steel—and the largest producer of CO_2. . . . In China, pollution has been moved from east to west and from the city to the rural areas. The rich consume and the poor suffer the pollution.

—PAN YUE, PRC DEPUTY MINISTER OF THE ENVIRONMENT, 2006

As industrial production has grown, so has pollution. You hang your clothes outside to dry, and they turn black!

—FUTIAN SHOPKEEPER

I MADE MY first trip to Futian in 2001 to explore whether the township would be an appropriate place to study the environmental consequences of small-scale industrial development in the countryside. Although it was late August, the summer monsoons were still in full force; on the way out to the township from the municipal center of Panzhihua, we had to stop in several places where the standing water on the road reached the undercarriage of our car and was too deep to pass. Once in Futian, Little Hu, in the company of several other cadres, gave me a ride in the government Jeep up the steep hill to the industrial compound, skillfully steering along the rutted, muddy road to the coking plant. My field notes from that day mention that I conducted interviews with several workers at the coking plant while standing on top of one of the smoldering brick furnaces, covering my mouth and nose with a bandana. Between the stifling heat, the monsoon humidity, and the acrid smell of coal smoke, I recall little beyond the cursory details in those early field notes, but I do remember feeling short of breath and nauseated. I also recall a small act of kindness that did not make it into my field notes: Mr. Zhang's wife, one of the investors in the zinc smelter, welcomed us out of the rain and into her house, where we spent a fair bit of the afternoon peeling and eating Asian pears and talking about her experiences as an industrial entrepreneur.

When my wife and I arrived in Futian in 2002 for a six-month period of field research, we did so with mixed emotions. I was passionate about my research, and my wife was gamely excited about the prospect of improving her Chinese and settling into the rhythm of life in the countryside. But we were also taken aback by some of the sensory details that a longer period of research allowed us to observe: the constant, lung-burning haze in the air; the factory laborers in soot-blackened clothes; the fine, gray dust covering vegetation, rooftops, laundry, and anything else left outside for more than a few hours; the endless, serpentine rows of coal trucks plodding along steep mountain roads; the local stream that often looked like flowing sludge. I will admit that both of us, overwhelmed by emotion, shed tears during the early days of fieldwork.

Researchers interested in environmental issues in rural China face some daunting problems. Relatively little information is available about local environmental quality in most townships and villages. Although the rural industrial sector is a major contributor to the national pollution burden, systematic air- and water-quality monitoring in the countryside is infrequent, sporadic, and incomplete at best. Furthermore, what little environmental-quality data exist tend to be aggregated into county or provincial reports, which makes it difficult to get any resolution at the township or village levels. As a baseline for understanding some of the environmental hazards posed by Futian's factories, part of my study involved conducting basic air-quality monitoring in the township. Some key components of China's environmental crisis stem from the nation's heavy reliance on coal as a primary industrial fuel source. The results of my air quality monitoring sessions, which focused on respirable particulate matter, or PM_{10}, allow us to begin to think about some of the threats to human health and the ecosystem in Futian and similar industrially dependent communities throughout China. Nevertheless, making a causal link between factory operations and pollution levels, or between pollution concentrations and adverse health effects for villagers, remains difficult and elusive. In part because of a lack of specific information about environmental quality, villagers' perceptions of factory pollution were marked by uncertainty and ambiguity.

THE SCOPE OF THE PROBLEM

Embattled by decades of neglect during the socialist period and decades of market-driven industrial growth during the reform era, China's environment is now at a crisis point. Nearly every scientist and government official can agree that major environmental problems—including deforestation, desertification, land degradation, and air and water pollution—threaten to undermine China's economic success. There is, however, little consensus about the root causes of these problems. In fact, socialist political systems have in general viewed environmental problems such as pollution as the byproducts of a corrupt capitalist mode of production, with its voracious appetite for profit and its tendency to externalize environmental and social costs. Socialism, by contrast, which ostensibly lacks a profit motive, is often seen as inherently less exploitative of natural resources. In this view, tacitly espoused by many Chinese leaders as well as academics, environmental degradation under China's socialist system is paradoxically an ideological impossibility and yet an undeniable material fact (Ho 2001). Many Western scholars harbor a similar romantic sentiment toward the socialist mode of production, even though China's period of high socialism was in fact marked by severe environmental degradation, including widespread deforestation and the proliferation of dams on rivers and streams (Shapiro 2001). The scale and pace of China's environmental crisis have undoubtedly been exacerbated by the past several decades of market-driven development; even conservative estimates suggest that pollution cuts 4 to 5 percent from China's GDP each year (Liu and Zhou 2001).[1]

Both a blessing and a curse, coal has fueled the spectacular rise of China's economy, which is currently the world's third largest (behind the United States and Japan) and is expected to become the largest within the next two decades. China is the world's leading producer and consumer of coal, gobbling up about 1.5 billion metric tons per year for industrial production as well as for household heating and cooking. About three-quarters of the nation's energy demands, which have grown fourfold during the reform era, are met by coal, with hydroelectric, oil and natural gas, and nuclear power making up the rest (see figure 4.1). China's coal stores, which amount to

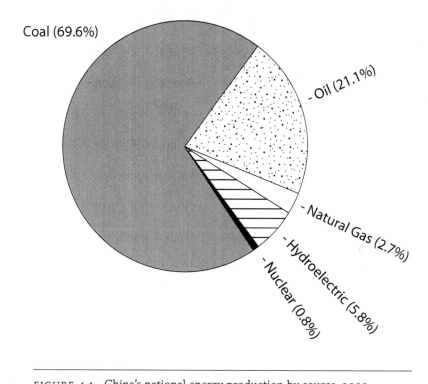

Coal (69.6%)

-Oil (21.1%)

- Natural Gas (2.7%)

- Hydroelectric (5.8%)

- Nuclear (0.8%)

FIGURE 4.1. China's national energy production by source, 2005.
Source: Adapted from Rosen and Houser 2007:17.

more than 100 billion metric tons, are so vast that it could keep burning at current rates without exhausting its supply until near the end of this century. It also increasingly imports coal from Indonesia, Australia, and elsewhere. Needless to say, the status quo provides little tangible incentive for the nation to diversify its energy sources.[2] Coal combustion is a major contributor to ambient air pollution and acid rain across much of the country, casting a dim haze over every major city. On commercial airline flights, passengers frequently gasp when, upon descent, they see the brown miasma that awaits them below. On one occasion, while touring Beijing's historic Tiananmen Square, I had to stifle laughter when I overheard one foreign tourist commenting to her guide that "Beijing's skies always seem cloudy."

China's industrial emissions increasingly affect the environment at the global scale. Pollution from northeast China, for example, follows the prevailing wind patterns, sullying the skies over Korea and Japan. Using organic chemical analyses, scientists in the United States are now able to determine with some certainty that pollutants from China (including carcinogenic compounds known as polycyclic aromatic hydrocarbons, or PAHs) are carried into the upper atmosphere, where they travel across the Pacific as far as the west coast of North America, contributing to the ambient pollution burden in Oregon, where I live (Primbs et al. 2007). In 2007, China surpassed the United States to officially become the world's largest producer of carbon dioxide, emitting 6.2 billion tons of the key pollutant responsible for global climate change (Vidal and Adam 2007).[3]

RURAL INDUSTRY AS CULPRIT

There are many reasons for China's dismal air quality. Large-scale factories, some owned by private or foreign investors and some still under state control, emit millions of tons of pollutants into the atmosphere. China's rising middle class, moreover, is increasingly enamored with the automobile, which means that nonpoint source pollution from car emissions is also a growing problem. But rural factories like those in Futian, which number in the millions, are also a significant and comparatively understudied contributor to China's overall pollution burden. Both Chinese and international scientists acknowledge the daunting scale of the problem. In a hefty tome dealing with the overall state of China's environment entitled *Case Studies in Environmental Protection*, the environmental scientist Meng Lang cites five major problems threatening the long-term survival of rural industry in China, four of which are directly related to the environment:

1. The rate of natural resource consumption in the rural industrial sector is unsustainably high
2. Environmental contamination from rural industry is increasingly pressing
3. Rural industrial pollution threatens the health and welfare of local residents
4. Rural industrial pollution damages industrial infrastructure as well as agricultural products
5. Unregulated growth in the rural industrial sector threatens the integrity of the socialist planned economy

(MENG 1999)

A number of factors contribute to this grim assessment. Because of its abundance and its state-regulated pricing structure, coal is almost universally the fuel of choice for rural industry. Factories located in rural areas often lack access to significant capital; as a result, they typically employ inferior, outdated technology and take few environmental-mitigation measures. China's overall industrial energy efficiency, which is a measure of the energy input required to produce a given industrial output, is approximately half the average for developed countries (Liu and Diamond 2005:1180); energy efficiency in rural industry is particularly poor. Modern coal-fired industrial facilities in the United States and Europe are typically equipped with "coal-scrubbing" technologies such as electrostatic precipitators, which remove much of the particulate matter from the gaseous emissions. But this is not true for most of rural China, where the majority of coal is burned unwashed; coal from the southern and southwestern regions, including Sichuan, is notoriously high in sulfur content. Sulfur dioxide and nitrous oxide are the two key byproducts of coal combustion that contribute to acid rain across much of the country.

Institutional factors also play a role. Several years ago, while chatting with an American toxicologist friend of mine, I described China's rural industrialization process and some of the environmental challenges it raised. After listening with some interest for a few minutes, he asked, "How do you regulate, inspect, and improve millions of rural factories?" His question hit upon a central problem related to the institutional capacity of China's environmental-oversight bureaucracy. The State Environmental Protection Administration, the agency charged with setting national pollution standards and enforcing compliance, has recognized the seriousness of rural pollution since at least the mid-1990s. SEPA has issued a series of policies and directives to control emissions, but the rural industrial sector remains a serious contributor to the national pollution problem.[4]

I had the chance to interact with several SEPA officials and quite a few scientists and regulatory officials in the Renhe District Environmental Protection Bureau, which had jurisdiction over Futian's factories. Although I generally found these individuals to be capable and dedicated professionals, most of them conceded that monitoring activities in Futian had in fact been extremely limited over the years. The Renhe District EPB had fourteen townships and more than 120 industrial factories under its charge; with limited manpower and technological capabilities, officials could conduct only sporadic monitoring and often felt overwhelmed by the scale of their responsibilities.

As I reviewed SEPA publications and statistics, I found this to be the case for much of rural China; despite the significant role of rural industry in China's air-pollution problem, little systematic monitoring takes place. I learned, for example, that Sichuan province emitted an estimated 1.3 million tons of particulate matter from industrial sources in 2000, but this gave me little clue about levels of exposure in Futian and the associated health outcomes (Sichuan Statistical Bureau 2002). In my interviews with EPB scientists and with factory managers, I was able to view the records of one air-quality-monitoring session that had taken place in Futian in late 1999 and that showed that the zinc smelter in particular was in violation of sulfur dioxide standards. This represented the only systematic environmental assessment in the township in recent years.

COPING WITH COAL

Particulate matter includes any substance other than water in the atmosphere with a size greater than molecular dimensions. Most often, PM appears in the ambient atmosphere as aggregates of various molecules and substances, including sulfate, organic carbon, nitrate, and elemental carbon. The sizes of particles within the PM mixture may vary greatly, from 10^{-4} to 10^4 micrometers. This is significant, since fine particulate matter, which can be inhaled deeply into the alveoli of the lungs and readily absorbed, is most closely linked with health problems such as bronchitis and asthma.

Particulate matter is admittedly an extremely coarse measure of environmental contamination, but it provides a useful proxy for understanding some of the pollution-related hazards with which much of the Chinese population copes on an ongoing basis. There is mounting epidemiological evidence that inhalable particulate matter—that is, particulate matter smaller than ten micrometers in diameter, or PM_{10}—is linked to increased rates of respiratory diseases, cardiopulmonary damage, and mortality (Samet et al. 2000; Hoek et al. 1997; Dockery et al. 1993).[5] One pioneering study conducted in six cities across the United States provides a good estimation of the health effects of PM_{10}. The results showed a 0.8 percent increased mortality rate for each ten additional micrograms per cubic meter ($\mu g/m^3$) of annual increase in PM_{10} exposure. As observed levels of particulate matter increased, so did observed levels of mortality (Dockery et al. 1993). Particulate matter often also coexists with other key pollutants such as sulfur dioxide (SO_2) and nitrogen dioxide (NO_2), and these pollutants bring with them their own deleterious health outcomes. It is extremely difficult for researchers to isolate the effects of a single pollutant within the ambient air mixture, since there are often interactive and synergistic effects between pollutants.

In China, the annual death toll from air pollution, primarily from industrial and household coal combustion, is estimated at 300,000 people (Economy 2004:19). However, epidemiological research on the effects of air pollution is in its infancy and tends to focus mostly on urban areas, where researchers have elucidated the link between particulate matter and daily mortality rates (Kan and Chen 2003; Xu et al. 2000); asthma and other respiratory ailments in children (Qian et al. 2000); and other health outcomes (Peng et al. 2002). In addition to its toll on human health, particulate matter causes significant health-related economic damage; hospital visits caused by the health effects described above, coupled with lost worker productivity and deterioration of infrastructure, are taking a mounting toll in China.

Perhaps the most difficult to adequately understand are the effects of particulate matter on vegetation and general ecosystem health. Vegetation is affected by particulates through various processes, including dry and wet deposition on plants and effects on soil chemistry. The deposition of PM in soils causes acidification, which can damage forest and agrarian ecosystems and also leads to the acidification of surface water and groundwater. These processes are complex and still poorly understood; the exact response of ecosystems and vegetation types to a given concentration of PM depends upon the particular size and chemical mix of the PM involved, as well as atmospheric conditions (Grantz, Garner, and Johnson 2003). Nevertheless, evidence suggests that PM may induce a phytotoxic response in vegetation, which can be devastating to crops and agrarian ecosystems, as well as to the people who depend upon these ecosystems for their livelihoods.

As scientists around the world began to uncover stronger evidence of the linkages among particulate matter, human health, and ecosystem integrity, most industrialized countries established domestic air-quality standards. At the international level, a World Health Organization working group established a set of air-quality guidelines in the 1980s based on available scientific evidence about the health effects of air pollution. The general consensus from this body of work, which is based on increasingly sophisticated toxicological and epidemiological methods, is that adverse health outcomes are seen at lower and lower pollution concentrations. Since the WHO is not a regulatory agency, its guidelines have no direct policy implications; countries must set their own regulations and standards to protect the environment and the health and well-being of their citizens.

In China, SEPA sets primary, secondary, and tertiary air-quality standards for key pollutants based on geographic zoning. Class I areas, which have the most stringent emissions regulations, are tourist, historic, and conservation areas; Class II are residential urban and rural areas, including agricultural areas such as Futian; and Class III are industrial areas and heavy traffic areas (Wan 1999; Chinese State Environmental Protection Administration 1996). Each class corresponds to a standard (*biaozhun*) for various key pollutants. China sets annual standards for a number of pollutants associated with coal combustion, including total suspended particulates (TSP), PM_{10}, sulfur dioxide, and nitrogen dioxide. These standards are listed along with WHO guidelines in table 4.1. Setting annual standards for pollutants means that the agency has established, based on varying degrees of scientific evidence, how much of a given pollutant is "safe" to breathe on a regular, ongoing basis. WHO guidelines for PM_{10}, for example, are based on the lowest levels at which total mortality, and mortality specifically related to cardiopulmonary problems and lung cancer, have been shown to increase with more than 95 percent statistical confidence in response to long-term exposure to this pollutant.

As table 4.1 shows, Chinese standards for these key pollutants are somewhat more lax than but generally in line with WHO standards. At a national level, China does not lack adequate scientific standards for air quality.

TABLE 4.1   SEPA standards and WHO guidelines for selected pollutants

Pollutant	SEPA standards (annual exposure levels) Unit = micrograms/m³			WHO guidelines
	Class I	*Class II*	*Class III*	—
Total suspended particulates (TSP)	80	200	300	—
PM_{10}	40	100	150	20
SO_2	20	60	100	20*
NO_2	40	40	80	40

* Twenty-four-hour average guideline. Guidelines for annual concentrations have not been determined.

Source: 1996 SEPA standards and Wan 1999.

However, implementing and enforcing these standards often falls to the Environmental Protection Bureaus, which vary in terms of technological, personnel, and institutional capacity. In rural areas in particular, scientists know little about current levels of these key pollutants or about the likelihood or severity of human health effects.

MONITORING PARTICULATE MATTER IN FUTIAN

Let me qualify what follows by reiterating the fact that I am an anthropologist, not an air-quality scientist. I entered the field armed only with some basic equipment and a modicum of laboratory training at the Northwest Center for Particulate Matter and Health in Seattle. In the lab, I was able to familiarize myself with the equipment, which included a personal monitoring air pump, a ten-micrometer particle size selector, and a few dozen Teflon filters. The basic idea behind PM sampling is quite simple. The technician runs the electrical air pump at a set rate of flow for a set period of time. Air is drawn through a length of latex surgical tubing, at the end of which is a particle size selector, which ensures that only PM_{10} is collected. The particles are drawn through the size selector and onto the Teflon filter, where they stick. Prior to conducting the sampling, the technician has weighed each filter down to the milligram on a microbalance scale. After sampling, the technician weighs the filter again (a process called "gravimetric analysis"); any mass gained between the first and second weighing sessions is attributable to particulate matter collected during sampling. A few simple calculations based on the rate of air flow and the duration of sampling give the technician a measurement of the concentration of PM_{10} in the air.

As part of my training, I had set up the equipment outside my home in Seattle for a seven-day trial sampling period. When they come out of the box, the Teflon filters are brilliant white in color. After the seven-day trial period in Seattle, as I lifted the filter out of the size selector and into a Petri dish with a pair of tweezers, I noted that it had darkened several shades to an ivory color. My training complete, I boxed up the equipment and didn't think about it again until I was in Futian in 2002.

After arriving in Futian, I set up the equipment outside our rented room at Li Jiejie's house in the open district, about 500 meters south of the industrial compound. I chose this location because it represented the hub of the township, where most stores, residences, and human activities were centered. On a more pragmatic level, it was also the only place in the township with a reliable source of uninterrupted electricity needed to operate the pump. After calibrating the pump and setting it to the correct flow rate, I turned it on and let it run, intending to wait seven days before removing the filter, as I had done during my training.

Outside, I could hear it humming along as I went about other activities. Then, approximately thirty-six hours after I had turned it on, the pump abruptly stopped. I couldn't figure out what had happened. I checked the electrical supply, which seemed fine. I checked the log on the pump's electronic display to see whether any curious villagers had tampered with it or changed the settings; they hadn't. Finally, I decided to open up the particle size selector and extract the filter. When I did, the problem immediately presented itself: the filter, now completely black from a heavy load of particulate matter, had overloaded the pump, which could no longer draw air through it. In a matter of hours, Futian's air had clogged the pump. The high levels of PM would force me to make many adjustments to my sampling protocol, including changing out the filters regularly during each seven-day sampling period.

After working out many similar problems, I was able to conduct four monitoring sessions, each seven days in length, between December 2002 and April 2003.[6] Table 4.2 provides a summary of observed PM_{10} concentrations during the four sampling periods. A time-series chart of the same

TABLE 4.2 Observed concentration of PM_{10} in Futian

Sample period	PM_{10} concentration ($\mu g/m^3$)	Factory activity
Sample period 1 (Dec 22–29, 2002)	247.6[a]	Only coking plant operating
Sample period 2 (Jan 15–23, 2003)	356.4	All factories closed
Sample period 3 (Mar 1–8, 2003)	114.7	All factories closed
Sample period 4 (Apr 20–24, 2003)	104.7 [a]	Only zinc smelter operating
Average of all sample periods	205.9	

[a] Difference between averages is significant, $p < .01$.

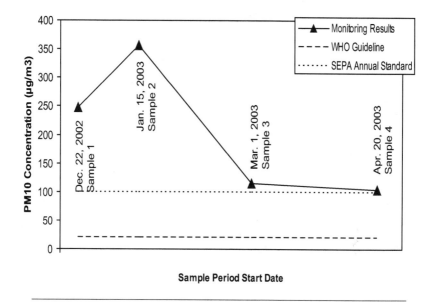

FIGURE 4.2 Time-series chart of PM$_{10}$ concentration in Futian (µg/m³).

data is shown in figure 4.2. Laboratory analyses, which were conducted at the PM Center in Seattle, revealed some interesting results. The average concentration for the sampling periods was 205.9 micrograms per cubic meter, 2.1 times higher than the SEPA annual standard for Class II areas and roughly 10 times higher than WHO guidelines. Accordingly, it can be expected that Futian's residents are susceptible to the kinds of health risks entailed by chronic exposure to PM$_{10}$, including cardiovascular and pulmonary diseases. In addition, the observed levels of PM$_{10}$ are likely to have a severe effect on vegetation, including crops, and on soil chemistry.

As part of my analysis, I attempted to correlate the observed levels of PM$_{10}$ with the factories operating in Futian during the sampling periods. Strangely, I found that there was no simple relationship between the two, as table 4.2 shows. The highest concentrations of PM$_{10}$ were recorded during the second sample period in January 2003; at this point, all of the township factories had been ordered closed by the EPB, an event that I describe in detail in subsequent chapters. The second-highest concentrations were recorded in December 2002, when only the coking plant was in operation. But the two periods during which any factory activity took place differed markedly in their ambient concentrations of PM$_{10}$. Sample period 1, during which the coking plant was the only factory in operation, showed a concentration of 247.6 µg/m³, while sample period 4, during which the zinc smelter was the only factory in operation, had a concentration level of only 104.7 µg/m³. Basic statistical testing revealed that the difference in concentration between these two periods was significant.[7]

Although it is impossible to determine the precise effect of any given factory on PM_{10} levels in the local ambient atmosphere, and although seasonal factors such as wind direction and domestic coal combustion for heating complicate the analysis, this result in part reflects the importance of environmental-mitigation technology. Both the zinc smelter and the coking plant, under orders from the Renhe EPB, had installed tall smokestacks in order to disperse pollutants higher into the air and mitigate potential health effects for workers. But the coking plant's smokestack had been badly damaged by an explosion in 2001, and factory investors had failed to repair it, citing a lack of capital. Coking-plant emissions were vented from open coke ovens directly into the air. Thus, while other factors no doubt influenced the pattern of PM_{10} concentrations that I observed, it is likely that part of the answer lies in the use of basic environmental-mitigation technologies.

The lack of a clear correlation between local factory activity and ambient levels of PM_{10} also raises a number of interesting issues to consider. It reflects, first of all, the centrality of problems of scale in air-quality monitoring. In particular, "transboundary emissions" have a known confounding effect on attempts to measure local air quality, and Futian illustrates this. The township is located within a nexus of rural industrial activity, with three other townships (Longdong, Daxing, and Shilongba), each with coal-burning factories, located less than ten kilometers away. Furthermore, Panzhihua Iron and Steel Corporation, the third-largest state-owned steel smelter in China, is located just fifty kilometers to the east and casts a heavy shadow over the entire region's air quality. Quite simply, air quality is a regional phenomenon; while local point sources may be the most obvious culprits, they are by no means solely responsible for local levels of PM_{10}.

RISK ASSESSMENT AND UNCERTAINTY

My foray into environmental monitoring taught me several lessons about the limitations of this emerging science, some of which preclude making broader interpretations from the data that I gathered. One pitfall with assessing health risks from air pollution by using PM data alone is that it is difficult to identify the independent effects of individual pollutants, given the variability of air mixtures. In short, without a detailed chemical analysis of PM_{10}, it is difficult to predict specific health outcomes (Brook, Dann, and Burnett 1997; Ito, Thurston, and Hayes 1993). An environmental toxicology laboratory is working on analyzing the Futian air samples for polycyclic aromatic hydrocarbons, a group of chemical compounds that are associated with coal combustion, many of which are known carcinogens. Preliminary analyses have suggested that PAH levels in Futian are elevated beyond established safety thresholds associated with risks to human health.

Another confounding problem is that ambient levels of PM_{10} do not necessarily reflect individual exposure to pollution on the part of workers or villagers, who, because they lived in different locations and went about different daily activities, were likely exposed to different concentrations of toxins. A quantitative epidemiological risk assessment would be a logical next step in determining the specific health threats villagers faced. The process through which risk-assessment experts (who may be environmental chemists, toxicologists, or public-health practitioners) determine whether a population is at risk from contamination is exceedingly complex. Scientists track a hazardous substance as it leaves a source, such as an industrial facility. They observe the substance as it moves through a medium, such as the air, and results in an exposure through respiration. This exposure, combined with the length of time a person is exposed, creates a "dose," which may result in a certain health outcome, or "response," in certain individuals.

But the risk-assessment process is imbued at every step with great scientific ambiguity (Tesh 2000:25), which makes it incredibly difficult to observe the dose-response relationship with much validity. First of all, direct experimentation on human subjects would yield the most valid and reliable results on the health outcomes of a given toxic substance, but such experimentation is generally impossible for obvious ethical reasons: who would be willing to serve as a research subject in a study of a toxic contaminant of unknown severity? Using animal models in the laboratory can provide researchers with an opportunity for some controlled experimentation, but extrapolating the results to the case of humans is problematic since there is no guarantee that animal models accurately predict human health risks. Conducting in situ epidemiological studies, in which scientists observe certain health outcomes (increased incidence of certain cancers, for example), leaves us with a huge validity problem: although health data may show greater risk of disease for a study population and environmental monitoring data may show that a given toxic chemical exceeds established safety thresholds, the problem of correlating these two types of data, much less discovering a causal relationship between them, remains. Making this link involves studying a dizzying array of variables, from the chemical toxicity of a contaminant to the possible exposure pathways, the various media of exposure, the intensity and frequency of exposure, and the spatial and temporal concentration patterns of the contaminant.[8]

Nevertheless, the levels of ambient PM_{10} I observed in Futian are far in excess of accepted standards and consistent with increased risk of mortality, respiratory and cardiopulmonary diseases, and harmful effects on ecosystem integrity, including damage to vegetation. Rural air pollution in China from industrial activity is a serious issue that deserves careful scrutiny. The lack of longitudinal data on pollution levels in Futian during the reform period makes it impossible to see exactly how the industrialization process or the privatization of industry has affected environmental quality. Moreover, air pollution is but one of many environmental problems that villagers have to grapple with.

THE AMBIGUITY OF HARM

During an interview, a young couple who lived in the open district introduced me to their daughter. A smiling girl with bright eyes and pink cheeks, she sat on the floor of the family's restaurant, which catered to busloads of Chinese tourists on their way westward to the scenic mountain town of Lijiang in northwest Yunnan. Her mother called for her to come over and say hello to the "old foreigner [*laowai*]"; obediently, she crawled across the floor toward us and her mother scooped her up in her lap. I noticed that the girl's knees were scarred and calloused, as though crawling was her primary way of getting around, though she appeared to be about two years old. "She will be two next month," her mother said, noticing my gaze, "but she still can't walk." Both the girl's feet were malformed, and, according to her mother and father, she had difficulty speaking and performing tasks that other children her age could do. Brushing away tears, her mother explained that they were trying to save enough money to pay for orthopedic surgery on the girl's feet.

Although it is impossible to link the girl's condition to environmental contamination with any scientific validity, the couple was convinced that exposure to pollution during the mother's pregnancy had played a role. Arsenic, lead, and mercury—all byproducts of metal smelting and coal mining—are capable of passing through the placenta and, in chronically high doses, can cause the effects seen in the girl's case, as well as a host of other threats to fetal health such as low birth weight and spontaneous abortion. Arsenic, although found naturally in many parts of the world as an inorganic compound associated with igneous and sedimentary rocks, is also a common byproduct of anthropogenic activities such as mining and metal smelting. A known carcinogen, it is associated with cancers of the skin, lungs, bladder, prostate, and kidneys. Arsenic and other metals follow multiple exposure pathways, entering people's bodies through air, drinking water, or food grown in contaminated soil. Groundwater contamination was a likely culprit in this case, given that most villagers use individual wells or drink water from a central groundwater-distribution system sponsored by the township government.

Establishing a causal link, or even a correlative one, between the girl's physical and mental disabilities and her mother's prenatal exposure to environmental toxins is further complicated by the fact that there is so little environmental oversight and monitoring in communities like Futian. This is a scientific ambiguity that the couple will have to live with. As the anthropologist Melissa Checker has pointed out in her study of contamination and illness in an African American community, "but I know it's true" is the oft-repeated refrain of communities whose daily experience of contamination runs head-on into the shortcomings and uncertainties of scientific environmental assessment (Checker 2007). In the process of thinking about environmental oversight in China, I couldn't help making comparisons with the United States, where the National Environmental Policy Act, signed into law in 1970, serves as the foundation for much of the nation's environmental law and policy. Under NEPA, any major actions that may significantly affect environmental quality are required to undergo either a full Environmental Impact Statement, or an abbreviated waiver process called a "Finding of No Significant Impact." In either case, proper steps must be taken to identify and mitigate the impacts on plant and animal species, ecosystem services, and human health.

By comparison, China's Environmental Impact Assessment Law (Huanjing Yingxiang Pingjia Fa) was passed in 2002. It was put to the test in January 2005 when SEPA temporarily stalled thirty large-scale industrial projects, many of them with investment from municipal and provincial governments, for failure to conduct proper environmental-impact analyses.[9] But this legal and regulatory framework is still in its infancy; when Futian's factories were constructed in the 1980s, and when subsequent infrastructural improvements and privatization took place in the 1990s, there was no regulatory framework that required a front-end assessment of their associated environmental or human health risks.

This family's story is one of the most poignant examples of the perceived threats from pollution that I heard during the course of my fieldwork, but it is hardly unique (see Tilt 2006). Countless residents discussed with me the deleterious effects of pollution on individual, family, and community health, although there was often little consensus on the precise nature or severity of the health threat. Many parents feared that pollution from township factories was somehow related to their children's health problems, including developmental disorders, asthma, and difficulty breathing, as well as chronic coughs and colds. Ambiguity is a common theme running through many of the villagers' narratives about the ill effects of pollution on human health in the community since no systematic ecological or epidemiological studies have been conducted to assess the potential risk from pollution.[10]

Mrs. Zhou, a Dai minority woman who lived near Li Jiejie, worked part-time in the sanitation clinic (*weisheng zhan*) in the open district. Her six-year-old daughter often attended the English classes that my wife and I periodically taught in the large meeting room in the township government offices. Mrs. Zhou's main responsibilities consisted of treating common ailments such as colds, fevers, and diarrhea, as well as giving vaccines to school-age children, which made her somewhat unpopular with her daughter's friends. The walls of the sanitation clinic were lined with metal shelves that contained both common Western pharmaceuticals, such as pain relievers and cold medicine, and traditional Chinese medicine tablets, powders, and tinctures. While most of her duties were routine, Mrs. Zhou told me that factory workers and people who lived near the industrial compound often came to the clinic complaining of respiratory problems and difficulty breathing from acute exposure to coal smoke. "There's nothing really that we can do for them," she conceded. "If their condition is bad enough, we send them to the hospital in Huaping or Panzhihua, but usually they just recover here for a day or two and then go back to work."

I was often surprised in spite of myself at the sophisticated knowledge many villagers had regarding the health effects of toxic substances. Given that nearly all of the meat, grain, fruits, and vegetables consumed in the township are produced locally, many villagers were concerned about the accumulation of toxins, especially from metal smelting, in the food supply. One young man told me, for example, that "toxins get into the plants and animals—cows, pigs, ducks—and then people eat them. It's not healthy." Many villagers expressed particular concern about arsenic (*pishuang*), lead (*qian*), and mercury, which people called "liquid silver [*shuiyin*]." Although there was no systematic environmental education in Futian, and little media coverage of pollution incidents, many villagers expressed a rather sophisticated lay knowledge of the processes of bioaccumulation by which lipid-soluble toxins are stored in the body's fatty tissues and, over time, produce a toxic response. In a standardized survey, a majority of villagers agreed with the statement, "Eating animals that have been exposed to pollution is hazardous to people's health." Some villagers felt that air and water pollution presented cumulative health effects that ultimately compromised the longevity (*changshou*) of township residents. "People don't live as long as they used to," was a common refrain. Longevity remains an important cultural value in rural China, where elders typically live in joint families and enjoy considerable respect.

But villagers had little choice about what foods to consume since food production and distribution take place on an essentially local scale. In the open district in the center of the township, for example, the daily morning market sold meat, fruits, vegetables and a few prepared foods such as steamed buns, or *baozi*, beginning at six A.M. All of these foods were locally grown in soil that was coated in a plume of smoke—containing arsenic and lead—for two decades. Most of the agricultural households in the township grew a substantial portion of their own food and purchased a few specialty items such as fruits or peanuts at the central market. Their diet was seasonally dependent: when the peas were ripe, people ate peas for several weeks in a row; when melon season arrived in the spring, villagers ate honeydew, cantaloupe, and watermelon by the dozens. A typical meal in Futian, as in most of rural China, consisted of rice (*fan*) and meat or vegetable dishes (*cai*), nearly all of which were cultivated locally.

Quite often, villagers relied on their sensory perceptions to assess the severity of industrial pollution. Seeing or smelling coal smoke from factories is easy when the pollution comes from a proximate source. Once I had become familiar with the locations of the factories in the industrial compound, I could easily tell which ones were operating by the location of the steady plumes of smoke emanating from the hilltop. But pollution in general, and air pollution in particular, is by nature something that crosses boundaries. The anthropologist Mary Douglas, in her classic study of ritual purity and contamination among the Lele of Africa, aptly described pollution as "matter out of place" (Douglas 2002 [1966]).

Air pollution from small-scale factories in the nearby townships of Shilongba, Daxing, and Longdong, as well as from Panzhihua Iron and Steel, cast a constant haze over the valley. Even when local factories were not operating, the horizon was dotted with smokestacks and billowing plumes of smoke, which could be seen from most locations around the township. Li Jiejie and many of the other local women perennially complained that soot and coal smoke soiled the clean laundry before they had a chance to bring it in off the drying line, and they worried about the residue left behind after they had washed their dishes. Township cadres, many of whom chatted and read the newspaper at a cluster of low tables in the courtyard in front of their office building during lunchtime, could be seen wiping a layer of soot from the tabletops each day. Villagers obtained drinking water either from naturally occurring springs or from the township central distribution system, which drew its supply from a well drilled just a few hundred meters from the factory compound and a cluster of abandoned coal mines. Although all villagers scrupulously boiled their drinking water to kill microbes such as bacteria and parasites, the water still retained a distinctive metallic taste, which most people covered up by adding copious amounts of green tea leaves. Concerned about the toxicity of the water supply, my wife and I ordered shipments of purified water for our own consumption, which arrived periodically from Panzhihua in large plastic containers, but the villagers had access to no such luxuries.

5

POLLUTION, PERCEPTIONS, AND ENVIRONMENTAL VALUES

Even out here in the village, you can see the smoke coming from the factory compound.

—FUTIAN FARMER

IN THE introduction to this book, I raised a key point of scholarly debate: In the midst of a decades-long national push for economic development, do the Chinese people care about the environment? If China is to achieve a more sustainable pattern of industrial development, it will need to rely on technological advancements, to be sure. But any serious attempt at environmental oversight and enforcement ultimately relies upon public awareness of environmental problems and public support for changing course. Much of the research on global environmental values makes the "postmaterialism" assumption, namely, that environmental consciousness is only made possible by the satisfaction of material needs through economic development. While wealthier nations and individuals, having met their basic survival needs and then some, have the luxury of worrying about air and water pollution, people in the developing world remain so mired in the daily struggle for survival that environmental quality is nothing more than an afterthought at best (Dunlap and Mertig 1997; Inglehart 1997; Dunlap, Gallup, and Gallup 1993a). In his now-classic description of the "risk society" produced by the technological developments of modernization, the German sociologist Ulrich Beck typified this line of argument: "On the international scale it is emphatically true that material misery and blindness to hazards coincide. . . . For these people . . . chemical factories with their imposing pipes and tanks are symbols of success [whose threat] remains largely invisible" (Beck 1992:41–42).

What little empirical evidence exists on environmental consciousness in China seems to at least partly support this thesis. Some of the most important and widely cited studies on China's environment depict its citizenry in general, and its peasants in particular, as too poor to worry about deteriorating environmental conditions (Wheeler, Wang, and Dasgupta 2003; Lo and Leung 2000; Edmonds 1998:726). Somewhat troublingly, one recent study even found that two-thirds of respondents in a rural county had never even heard the term "environmental protection [*huanjing baohu*]" (Alford et al. 2002). The prevailing scholarly view thus envisions a Chinese peasantry fatalistically breathing polluted air and drinking contaminated water in the name of national economic development. Even Vaclav Smil, arguably the most accomplished scholar of the Chinese environment, makes this assumption; in *China's Past, China's Future*, a recent overview of the nation's environmental decline, he argues:

> In general, the attitudes of people who have just emerged from long years of privation to the threshold of life promising a bit more freedom and a little more prosperity are not conducive to conservation, savings, and the eschewing of immediate consumption; just the opposite is likely to be true, putting further accelerated pressure on the environment. Indeed, here is a perfect illustration of a key ecological concept well known as the tragedy of the commons, or killing the goose that lays the golden egg.
>
> (SMIL 2004:144)

When I began my research on environmental values in rural China, it struck me that the empirical basis for these claims was somewhat thin. The line of argument seemed loosely based on the "environmental Kuznets curve," which posits that environmental problems such as pollution tend to increase during the initial stages of economic development, when public and private resources are mobilized for industrialization, then decrease over time as a result of technological innovation, increased environmental awareness and regulation, and the externalization of pollution sources to poorer countries or regions (Grossman and Krueger 1995). In fact, scholars have done a fairly good job of testing this relationship between economic development and environmental degradation, and it seems to hold up well in a variety of empirical contexts. But applying similar logic to the study of people's environmental *values* requires two methodological and analytical leaps that make it difficult to interpret the results.

The first "leap" relates to the problem of scale. Most human-environment interactions occur on a local, rather than a global, scale. In the case of industrial pollution, for example, people may notice the effects on their health, their economic livelihoods, or their food safety. Their perceptions, for the most part, stem from direct experience of environmental contamination. Studying global environmental values requires the aggregation of data at macroscales such as the nation or region, which ultimately masks a great deal of the richness and variation in how people interact with their local environments.

Second, large-scale surveys often suffer from the difficulty of framing the issues in ways that are culturally, politically, and economically salient for the people questioned. It is my contention that a complete understanding of environmental values, as well as an understanding of how values shape actions and behaviors, requires in-depth studies of specific places. By examining the environmental values of villagers in Futian as expressed in their own narratives about the effects of industrial pollution on their lives and livelihoods, I wish to contribute to this debate. Rather than discuss environmental values in the abstract, my goal is to show how the villagers' ideas about their local environment were situated within their lived experiences. In particular, expressions of environmental values tended to be linked with concerns about the dramatic changes taking place in Futian's political economy, including the privatization of factories and the marginalization of farmers, who depended on an agro-ecosystem that was being undermined by industrial pollution.

THE HISTORIC AND PHILOSOPHICAL BASES OF ENVIRONMENTAL VALUES

Villagers' perceptions of environmental contamination were often situated in a historical context that included events that preceded Reform and Opening. During a casual conversation, Mayor Hu, the retired former mayor of the township, remarked,

> Compared to before, the environment has been totally destroyed [*quanbu pohuai le*]. In the past, you could hardly see through all the trees on the mountainside around here. When I was a youth in the 1950s, the mountains were totally covered in trees. They were this big around [making a circle with outstretched arms]! But people logged the mountainsides to get timber for building houses, and for fuel.

Mayor Hu insisted that environmental destruction was not endemic to the reform period; the Maoist era, too, had wrought its own environmental problems. Massive deforestation in the 1950s and 1960s, enacted in part to support Great Leap industrialization, had resulted in the loss of vast forest tracts and the erosion of mountainsides. However, the former mayor was careful to point out that Reform and Opening had introduced environmental hazards that were qualitatively different from what the villagers had experienced before: synthetic chemical fertilizers and pesticides designed to increase crop yields for the township's smallholder farmers; a series of small coal mines to extract the raw materials needed for Futian's industrial push; and, most significantly, the township's burgeoning factories.

As the historian Lynn White Jr. has observed, "Human ecology is deeply conditioned by beliefs about our nature and destiny" (White 1967:1205). Perceptions and values regarding the environment cannot be understood without an appreciation of the historical, philosophical, and cosmological trends that shape people's ideas. For example, the notion that nature exists "out there," someplace separate from human affairs, is the product of the Western philosophical tradition. As a basis for the preservationist ethic of someone like John Muir, the eminent American adventurer and author, this dualistic worldview underpins the idea of "wilderness" apart from human activity. Yet as the environmental historian William Cronon reminds us, "Wilderness is not quite what it seems. Far from being the one place on earth that stands apart from humanity, it is quite profoundly a human creation—indeed, the creation of very particular human cultures at very particular moments in human history" (Cronon 1995:69).

Wilderness, in other words, is an epistemological construction made possible by specific historical and cultural forces. One need look no further than the cornerstone of Western thought, the Bible, to find the human-nature dichotomy; the creation myth in the book of Genesis, for example, depicts a humanity divinely set apart from nature by God himself. In the creation story, humans are sometimes given dominion over nature—"Let us make man in our image, in our likeness, and let them rule over the fish of the sea and the birds of the air" (Genesis 1:26)—and sometimes expected to exercise righteous stewardship over God's precious resources. But they are always afforded a separate, superior status.

Curiously, this tendency toward a rigid separation between humans and nature in Western philosophical thought, when combined with modern ecological science and management practices, has resulted in a biocentrism akin to the American naturalist Aldo Leopold's widely cited land ethic: "A thing is right when it tends to preserve the integrity, stability, and beauty of the biotic community. It is wrong when it tends otherwise" (Leopold 1949:224).[1] This dualistic worldview is now evident in much of the First World environmentalism that drives global conservation efforts, which take the separation—and sometimes forced exclusion—of humans from nature as a precondition to conserving natural resources, threatened species, and biodiversity (West and Brockington 2006).

In contrast to the Western, biocentric understanding of the environment, traditional Chinese views about nature tended to treat humans as integral parts of the natural realm. For example, just as Confucianism has afforded Chinese society with a penchant for social order—sometimes rigidly, hierarchically so—so, too, has it provided the basis for an ecological worldview in which humans are inextricably linked with the natural world. The Confucian worldview is *anthropocosmic*: humans exist in relational resonance with, and indeed must maintain a delicate balance with, the natural forces of the universe. This relationship can be thought of as a triad consisting of heaven (*tian*), earth (*di*), and humans (*ren*) (see Tu 1998).

Taoism, too, emphasized harmony and continuity between humans and the natural world they inhabited. Consider chapter 78 in Laozi's enigmatic collection of wisdom, the *Tao Te Ching*:

Under heaven nothing is more soft and yielding than water.
Yet for attacking the solid and strong, nothing is better;
It has no equal.
The weak can overcome the strong;
The supple can overcome the stiff.

Here, one conjures up a mental image of falling water gradually wearing away bedrock over the course of millennia or of water droplets freezing in cracks of stone, expanding and cleaving the stone apart. The image of water, one of the softest and most fluid elements in nature, breaking down solid stone is a powerful one. Many aspects of Taoist thought seem contradictory, paradoxical, or downright confusing, chief among them the concept of *wuwei* ("inaction," or "not doing"), which conceives of the ideal relationship between humans and nature as one in which humans refrain from taking a dominating role and choose instead to simply exist in homeostatic balance with it. Indeed, carefully practiced, Taoism would reject altogether a clear disjuncture between nature and culture, just as the flowing, curved lines of the yin-yang symbol defy easy distinction (LaFargue 2001).

These are important philosophical traditions that undoubtedly play a tacit role in rural Chinese people's perceptions of the world around them and their place in it. Yet the influence of such abstractions about the environment is limited in places like Futian, for two reasons. First, throughout much of Chinese history, Confucianism and Taoism were the discourses of a somewhat elite literati. They were rich literary traditions that required a knowledge of written classical Chinese and a fair amount of leisure time to understand and practice. For rural people engaged in farming, interacting with nature—plowing the fields, irrigating, and harvesting crops—was a much more quotidian exercise that centered on daily toil. Second, the tumultuous forces of twentieth-century Chinese history diminished the impact of these traditions. Both Confucianism and Taoism were considered "feudal" customs under Maoist socialism and were thus severely restricted in favor of a more utilitarian ethic; under socialism, nature, like everything else, existed to "serve the people [*wei renmin fuwu*]."

EATING FROM THE MOUNTAIN

Mr. Li, a forty-eight-year-old Shuitian farmer, lived in Wuzitian with his wife and two school-age children on a narrow plot of land consisting of only three *fen* (three-tenths of a *mu*, or less than one-twentieth of an acre). He grew primarily rice and sweet potatoes. The latter he cultivated on a sloped corner of his land without irrigation, assiduously carrying buckets of water from the river each day to sprinkle over his crop. The family had faired poorly during the land reforms of the early 1980s; their meager land holdings barely supplied them with enough grain and vegetables for their own subsistence and a bit of surplus, which they sold in the daily market in the open district. Their cash income amounted to approximately 200 yuan per month, a sum that was spent almost entirely to support the educational expenses of the eldest son, who had scored well enough on the university entrance exam to attended Panzhihua University. The family's precarious economic conditions and their complete dependence on agriculture made them particularly vulnerable to the effects of pollution on the agro-ecosystem. Mr. Li's chief complaint focused on the coal-washing plant in Futian and a similar factory in the neighboring township of Daxing. Effluents from these factories, he complained, threatened him with financial ruin:

> We've had a terrible experience with the coal-washing plants in Daxing and Futian. The plants are upriver from here. Starting in about 1991 or 1992, at certain times of the year when the plant is running a lot, the whole river turns black. We don't want to irrigate our crops with it, and the animals can't drink from it.

I witnessed this occurrence several times myself. Without warning, the stream—which villagers called by different names, depending on which village it was running through—turned completely black for a period of days or even weeks. This tended to be when the coal-washing plants, which operated somewhat sporadically, were running at high capacity. As I described in chapter 3, these coal-washing plants remove some of the sulfur content from the raw coal through a mechanical process that involves using water to flush the coal down a conveyor belt and through a machine that agitates it, mixing the coal into a watery slurry. Workers collect the slurry in a large catch basin and allow it to dry then sell it to local households for heating and cooking. This process makes the coal burn cleaner, but it creates another problem: waste water. The effluents from the coal washing plants, essentially coal dust dissolved in water, are released untreated into the local stream.

Mr. Li's assessment of the pollution problem stemmed from direct experience of its effects on his livelihood; with such a narrow margin of economic viability, he was acutely aware of the negative effect that water pollution had on his crop yield and on his stock animals. Many villagers were also concerned about the dry deposition of particulate matter and soot on their crops from air pollution, primarily from the zinc smelter and coking plant. The soot, they maintained, left a grey residue on the plants, compromising photosynthesis, killing plants, and lowering their yields. In early 2000, Mr. Li, along with a thirty-three-year-old Han man from his production cooperative in Wuzitian, filed formal complaints with cadres in the Futian and Daxing Township governments. They also reported the case directly to the Renhe District Environmental Protection Bureau. With EPB officials acting as arbiters, a compromise was reached whereby the government of Daxing would pay fifty yuan to each of approximately one hundred farming households downstream in Futian who had been affected by effluents from the plant. Futian's coal washing plant, which at that time remained a township enterprise, was never fined.[2]

The sum could scarcely begin to cover the financial losses of many farmers, some of whom had lost a whole season's crop. Local farmers, given the pittance of the settlement, held this up as an example not of the triumph of local citizens organizing against pollution but as a case of bureaucratic snubbing of those without economic and political power. "Fifty yuan!" Mr. Li told me, rolling his eyes, "What are you going to do with fifty yuan? They ruin your crop and then they give you fifty yuan!" This sentiment was echoed by many other farming households who cultivated land along the stream in Wuzitian village. Long after these formal complaints had been addressed, the pollution continued unabated, and many local farmers remained resentful of the factory owners and workers.

The farmers' expressions of environmental awareness must also be read alongside the broader political and economic changes that were taking place in the township. In chapter 3 I described how local factories were transformed from collective assets that provided revenue for community development into privately held companies that enriched a select group of investors but marginalized local community members. That transition became linked in the minds of villagers with the environmental degradation caused by the factories, although it is likely that pollution levels were also high before factory privatization. Futian's zinc smelter, coking plant, and coal-washing plant were all purchased by private investors, most of whom were urban people with considerable financial resources, and all of whom hailed from outside the township. In this way, Futian has experienced the "dismantling of communal capital" (Muldavin 1996) endemic to reform-era China.

In most cases, these outside investors brought with them their own labor forces or culled from the ubiquitous stocks of migrant labor throughout the countryside. The privatization of local industry was thus compounded by the outsourcing of labor opportunities, leaving local villagers dependent primarily upon subsistence and small-scale market agriculture. Where many households had at least one member who earned wages from local factories during the period of collective ownership, after the privatization of industry these households lost a valuable source of cash income. Villagers used the colloquial, mildly derogatory term *"waidiren* [outsiders]" to refer to the migrant laborers and recently arrived investors, and the memory of communal benefits from industry gradually faded with the completion of the privatization process. As one farmer told me, "[Industrial development] has brought a lot of *waidiren* into the community to invest in industry, but we don't get any of the benefits from industrial development."

In monetary terms, the benefits of industry were considerable. According to my survey, households with at least one family member working in industry earned on average 2,043 yuan per month, while households without any income from industry earned just 801 yuan. Households that relied solely upon farming for their livelihood constituted the majority of households in the township and earned, on average, just 433 yuan per month, a sum that was supplemented with in-kind income from farming. By contrast, the private investors in Futian's factories earned as much as 10,000 yuan per month, an unimaginable sum for most local residents. The agricultural system of Futian, although it comprised the bulk of employment and subsistence for the township's residents, was at constant risk of being undermined by local industrial activities. In the minds of most villagers, the agricultural livelihood was capable of raising living standards only so much, and many agricultural households in the community were aware of their tenuous position within the regional political economy. As one man put it, "Agriculture solves the problem of warmth and fullness [*wenbao wenti*], but it doesn't make you well off."

Villagers' concerns about environmental contamination were grounded less in abstractions about aesthetics and harmony and more in a sense that their economic livelihoods were being threatened. In their conversations with me about the environment, many villagers referenced an old saying common in rural China: "When you live on the mountain, you eat from the mountain; when you live near the water, you drink the water [*kao shan chi shan, kao shui chi shui*]." Land is the villager's lifeblood. In addition to being the source of day-to-day sustenance, it also constitutes the only real social security most farmers will ever know. In the wake of liberal reforms, farmers are exposed to considerable economic risks; no longer subject to the communal land system, they must meet their own economic needs as the state provides less security and fewer services than during the socialist period.

This problem is exacerbated somewhat by China's complex land-tenure system, which is highly variable from region to region but is commonly seen to contain four fundamental dimensions: the right to residual income from agricultural activity; the right to use land in relative freedom from state regulation and other encumbrances; security of tenure rights into the future; and land-transfer rights (Liu, Carter, and Yao 1998). In Futian, as in other areas throughout rural China, land rights are vested in the rural collectives. Farmers are given certificates that grant them use rights, but not full ownership rights, over their land plots. The central government shows no signs of lessening its commitment to collective land ownership and long-term land leases under the Household Responsibility System. In practice, this serves to maintain a smallholder agricultural system, where farmers intensively cultivate on average only three *mu* (0.2 hectares) of land and have little chance to participate in an agricultural economy of scale. Their economic livelihoods are thus only as good as their small plots of land allow, and a failed cropping season can mean disaster. In this way, Reform and Opening brought with it the individualization of two kinds of risk, each intertwined with the other: the economic risk of household-based farming, with its narrow and unpredictable profit margins; and the environmental risk of factories that generate profits for private investors while externalizing environmental damage onto the community at large. An observation made by the eminent Chinese anthropologist Fei Xiaotong more than a half century ago now seems particularly prescient: "Only those who make a living from the land can understand the value of land" (Fei 1992 [1947]).

MARGINALIZATION, KNOWLEDGE, AND TRUST

In addition to experiencing directly the negative effects of industrial pollution, local farming households were engaged in a constant struggle with the township government over its prioritization of industry in local development plans. The township government, villagers complained, constantly emphasized industrial development while downplaying the importance of agriculture, despite the fact that a majority of households in Futian depended solely upon agriculture for their livelihoods. The epitome of this privileging of industry was the construction of a 2,000-kilowatt hydroelectric power station below the township's main reservoir, which cadres initiated with an investment of 1.5 million yuan.

Township cadres pursued this course for two reasons. First, a decreased market for coke at Panzhihua Iron and Steel Company, coupled with a gradual depletion of local coal reserves, meant that many of Futian's factories were on shaky financial ground, making the prospect of future industrial profits and taxes uncertain. Second, officials from the Renhe District Environmental Protection Bureau were discussing the possibility of shutting down Futian's factories for noncompliance with emissions standards, a decision that would have far-reaching economic consequences for the township government (discussed in the next chapter). Local cadres planned to generate power year-round at the hydroelectric station and sell the surplus electricity to the neighboring township of Shilongba, located across the provincial border in Yunnan. This plan was laid out in township government documents and discussed at length in government meetings, where cadres estimated a potential return of 1 million yuan per year. Construction was underway throughout 2002 and 2003.

Mr. Zhang Huachao, the Communist Party secretary of Jingui village and an old classmate of Li Jiejie's, discussed the project with me. We sat in the main room, or *zhengfang*, of his house, peeling and eating mangoes, which he grew in a small field adjacent to his house. On the far wall was a strange amalgamation of décor: his certificate of membership in the CCP; a poster of Guanyin, the Buddhist goddess of mercy; a socialist-realist portrait of Mao Zedong, Deng Xiaoping, and Zhou Enlai gazing triumphantly into the distance; and a poster featuring a fat Maitreya Buddha with the Chinese characters for good luck and prosperity. In a prominent position on the center of the wall was a black and white television set, which was connected by a cable to the family's satellite dish, which he had recently installed in the courtyard near the donkey pens. Party Secretary Zhang's children—a girl who attended high school in Panzhihua, and a boy, an unusually tall junior-middle-school student—often complained that although they could view a wide variety of channels, they could not watch their favorite programs in color. As a result, he told me, a color TV set would likely be their next major cash expenditure.

In Party Secretary Zhang's view, Jingui farmers bore a disproportionate share of the costs related to the township's hydroelectric project. During construction, water had to be diverted from the main river channel, and rice paddies in Jingui were completely cut off from any irrigation source. It also became increasingly clear to villagers that even after the hydroelectric station was completed, irrigation water would be scarce in Jingui since a major share of the river water would be permanently diverted in order to turn the turbines and generate electrical power. Water shortages affected most households in the village. Because of its distance from the main road and its fortuitous topography, which was considerably flatter than the other villages in the township, Jingui enjoyed comparatively large land plots; many households cultivated up to five *mu* of irrigated rice paddy (*tian*) and ten *mu* of dry fields (*di*). However, villagers hadn't been able to plant a winter wheat crop in 2002 for lack of water. By early spring of 2003, it became clear that the power station would not be completed in time to supply the village with irrigation water, and villagers fulminated about the likelihood of losing their main summer rice crop.

Many villagers in Jingui, emboldened by the anonymous nature of my interviews with them, complained bitterly about their grain losses, which they viewed as a calculated privileging of industry on the part of local cadres. One farmer in Party Secretary Zhang's production cooperative complained that he had been forced to purchase rice for his family, something he never envisioned having to do in this fertile corner of the township. Others even went so far as to insist that I show the results of my interviews and household surveys to township cadres so that the full measure of the villagers' disaffection might finally sink in.

In the eyes of most farmers, industrial development was the culprit. By insisting upon promoting industry over agriculture, despite the fact that a majority of Futian's households depended primarily on agriculture for their livelihoods, cadres were sacrificing the well-being of the "old hundred names [*laobaixing*]." Factories provided a huge share of the township's tax revenue and also supported cadre salaries. It was no secret that township cadres' yearly performance evaluations (*kaohe zhibiao*), which determined their future salaries, promotion opportunities, and other benefits,

FIGURE 5.1 Rice fields could not be irrigated in Jingui Village after cadres diverted the river for hydropower development.

emphasized industrial growth. As a result, industrial laborers and investors wielded substantial influence over government officials and over the agenda of economic development in Futian, often at the direct expense of local farmers. Their powerful position within the local political economy and their only cursory reliance on the local agrarian ecosystem assured laborers and managers that their economic needs would be met while the environmental risks of industry were externalized to others in the township.

For all their animosity toward the outsiders (*waidiren*) who had exploited business and employment opportunities in Futian and profited from polluting activities, local residents in Futian also ultimately criticized what they perceived as the unfair development policies and practices of the government. They expressed a deep-seated mistrust toward both local cadres and district-government officials, who were supposed to play, in the eyes of local residents, the dual roles of development promoter and public-health protector.[3] The power to control and disseminate information about the local pollution problem was a central concern for farming households, who questioned whether the township government had the best interests of the common people at heart. As one villager told me:

The government is supposed to tell you when the pollution is bad [so you can take steps to avoid exposure]. The small-scale factories have grown so fast, with very little regulation. I have to say that these factories have failed to do what the central government wanted them to do, which was to improve the standard of living in the countryside. The problem is, they're all run by outsiders [*waidiren*]. All the taxes—and maybe bribes, too—go to the township government, and the old hundred names [*laobaixing*] get nothing. But they don't care about the interests of the old hundred names. All they care about is more investment and more money.

This stands in contrast to other government slogans and campaigns that cadres initiated to instruct villagers on various matters of public interest, including hygiene and disease prevention, family planning, and similar matters. Some environmental issues also garnered the attention of township cadres and became the objects of public campaigns. One such campaign was designed by the township forestry bureau to help educate the villagers about preventing forest fires, which could be disastrous in this relatively arid region. It was common for cadres to hold educational sessions regarding the controlled use of fire to burn away unwanted crop residues left in the fields after harvest, or to paint pithy slogans in public places where they could be seen by most villagers. For examples of such public campaigns, see figures 5.2 and 5.3.

Many farmers believed that local cadres knowingly turned a blind eye to industrial pollution in Futian because of the township's relatively weak position within the regional political economy and because of the ever-present need for township revenue. As one villager noted, "The government allows these polluting industries to exist in rural areas because they know people need the development and won't object." Curiously, the reform era has seen a realignment of political loyalties in many rural parts of China, where central policies such as Reform and Opening and the Household Responsibility System are seen as "upright [*zheng*]" and local cadres are seen as "crooked [*wai*]" (see Li 2004). This is likely a reflection of the fact that local cadres are now expected to be entrepreneurial, creatively "going toward the money [*xiang wei qian*]" by seeking out new sources of revenue to support the government budget, which leads many villagers to see them as predatory and self-interested. Villagers were aware of the environmental and health risks from factory smoke and water effluents, but, faced with a local government bureaucracy that privileged industrial development, they felt there was little they could do to protect themselves.

FIGURE 5.2 Public campaign slogan outside township government office: "Plant trees to establish the forest and protect the forest for the benefit of the nation and the people."

FIGURE 5.3 Public campaign slogan: "Protect the green garden, prevent forest fires!"

PATTERNS OF ENVIRONMENTAL CONCERN

In my discussions with villagers in Futian, it became clear that not all township residents were equally concerned about factory pollution. I subsequently became interested in several key questions: How do people in Futian define the locally salient risks from pollution? Do different groups perceive pollution in different ways? If so, what accounts for this variation? In an attempt to answer these questions, I interviewed a stratified random sample of informants from three different occupational groups: industrial workers, commercial and service-sector workers, and farmers. In the interviews, I asked informants about the costs and benefits of local industry; asked them to free-list the specific environmental and health risks posed by industrial pollution as they saw them; and encouraged them to discuss their perceptions with me. Analysis of these interviews revealed that seven "risk themes" stemming from industrial pollution were most salient for people in Futian; these themes, along with their frequencies of appearance within the interviews and a brief description of each, are shown in table 5.1. The themes represented a diverse set of concerns, most of which stemmed from direct experience in coping with pollution on a daily basis.

The next step in my analysis was to examine whether and how perceptions of industrial pollution differed across these occupational groups. I recruited another stratified random sample of 122 informants from the three different occupational groups, including industrial workers, commercial and service-sector workers, and farmers. This sample, which included both men and women and both Han and Shuitian ethnic groups in roughly equal proportion to their appearance in the township population, covered 17 percent of the approximately 700 households in the township.

I constructed a standardized survey that consisted of several sections, including demographic and socioeconomic information, views on local living standards, and perceptions of pollution from local factories. In the section on perceptions of pollution, I asked informants to indicate their level of agreement, on a scale of one to five, with a set of statements; each of these represented a risk theme drawn from the interviews. For example, the statement on the "direct health effects" theme was, "Pollution in Futian affects people's health." The survey provided me with scaled, ordinal data that allowed me to compare ratings across individuals and groups.[4]

To determine whether different stakeholder groups viewed the severity of pollution differently, I conducted a relatively simple statistical procedure called a proportional odds model. My working hypothesis was that the occupational groups would differ in terms of how they rated each risk theme;

TABLE 5.1 "Risk themes" related to industrial pollution cited by villagers

Risk theme	Informants using theme		Description
	(N)	(%)	
1. Direct health effects	34	94.4%	Pollution causes health problems for the informant, his/her family, or the community in general.
2. Damage to plants or crops	23	63.9%	Pollution damages agricultural crops and other plants.
3. Damage to visibility or scenery	23	63.9%	Air pollution impairs visibility or otherwise damages the aesthetic quality of the area.
4. Threats to animal health	14	38.9%	Water pollution contaminates drinking water for livestock.
5. Economic losses	8	22.2%	Air and water pollution damage crops, lowering yields and resulting in economic losses in the township.
6. Effects on food	5	13.9%	Eating plants and animals that have been contaminated by pollution poses a health threat for humans.
7. Threats to longevity	3	12%	Exposure to pollution limits peoples' ability to live a long life, an important cultural value in China.

conversely, my null hypothesis was that the groups' ratings would be identical. If the proportion of industrial workers who "strongly agreed" that pollution affects human health equaled the proportion of commercial and service-sector workers and farmers who also strongly agreed with this statement, then the response distributions of these stakeholder groups would be considered the same and the null hypothesis would not be rejected. The null hypothesis would be rejected only if the difference between the response distributions of any two occupational groups was larger than would be expected by random chance. I then ran contrast tests using the Chi-square statistic to see which groups differed in their responses at a statistically significant level.[5]

Figure 5.4 shows how informants in each group rated the severity of each risk theme. This is expressed by the percentage of responses from each occupational group that fell into each response category ("strongly agree," "agree," etc.). In this figure, the white portion of each bar reflects the percentage of informants in each group who strongly agreed with the statement, the lightly shaded portion the proportion of informants who agreed with the statement, and so on. A majority of my informants, regardless of which occupational group they belonged to, perceived industrial pollution as posing some risk to themselves and to the community. For most of the seven risk themes, more than half of informants "strongly agreed" or "agreed" with the statements.

However, the most interesting findings illustrated by figure 5.4 come from a comparison of ratings across occupational groups. Farmers and commercial and service-sector workers were quite similar in their responses, "strongly agreeing" or "agreeing" with the statements for most risk themes. Their responses on themes related specifically to the agro-ecosystem, including "theme 2: pollution damages plants or crops" and "theme 6: pollution affects the food chain," were unequivocal; about two-thirds of these informants strongly agreed that industrial pollution posed a significant threat. But the responses from industrial workers were especially striking in the degree to which they diverged from other groups; workers consistently provided ratings well below those of the other groups, and their responses were statistically different from the other groups on every risk theme. The percentage of industrial workers who "strongly agreed" with any risk theme was much smaller than for the other stakeholder groups, as illustrated by the comparatively short white portion of the bar chart for industrial workers across all risk themes. Furthermore, their lowest agreement ratings were for "theme 1: pollution affects people's health"; more than one-quarter of workers strongly disagreed with this assessment. This puzzled me, since most of

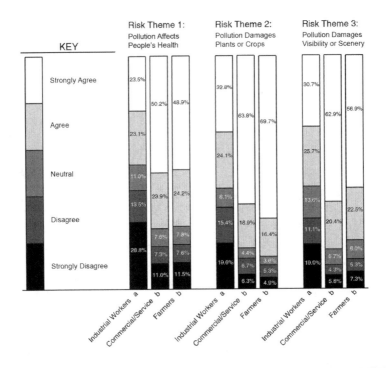

FIGURE 5.4 Occupational group ratings of risks from industrial pollution. *Note:* For each risk theme, different letters (a or b) indicate occupational groups whose response distributions differ at the p < .05 level of significance as determined by Chi-square contrast tests in the proportional odds analysis. For example, for risk theme 1: pollution affects peoples' health, industrial workers (a) differ in their responses from both commercial and service-sector workers (b) and farmers (b), p < .05. Commercial and service-sector workers and farmers do not show statistically different response distributions.

my surveys were conducted on-site at the factories, where workers toiled constantly in soot-blackened clothes while breathing toxic emissions.

EATING BITTERNESS

Near the factory compound, on the hillside of dry grass and scrub to the north of the open district, was a scatter of hastily built, one-room brick houses where the industrial laborers and their families lived. I visited one of

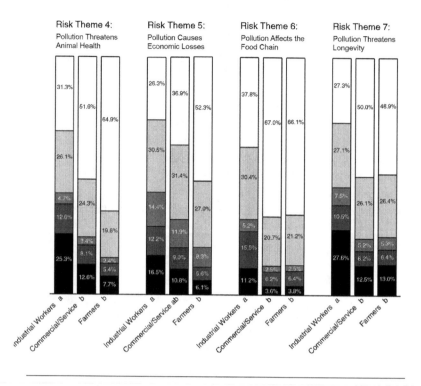

FIGURE 5.4 *(continued)*

the migrant workers, a Han man who lived inside the industrial compound with his wife and small son. The family had moved to Futian from their ancestral home in Guizhou in order to find work in the zinc smelter, where in a good month the young father could earn 800 yuan. The family's house, which consisted of a single room with a floor of packed earth, was less than a hundred meters from the zinc smelter, and the air smelled sharply of coal smoke. Shards from broken ceramic crucibles used in smelting the zinc ore littered the ground, which was covered in a thick layer of black soot. The young boy frequently played nearby in piles of discarded coal and zinc-ore slag.

My interactions with this particular family, and with other factory workers, helped to provide a framework for understanding why industrial workers' views of pollution diverged so sharply from those of others in the township. As we talked, I found that on the subject of industrial pollution both the worker and his wife were adamant that there was no harm to themselves or their young son. "The pollution has no effect on people's health," he insisted. "I've been doing this kind of work for years with no health problems." The couple's main concern was not exposure to toxic emissions; rather, they were worried that inspectors from the Renhe District Environmental Protection Bureau were visiting Futian with greater frequency, threatening to close local factories for noncompliance with air-pollution emissions standards. In their view, environmental regulations constituted the greatest threat to the township's well-being. As the young father told me, "They [EPB officials] say there's pollution, but it's not bad. Anyway, how can you develop without jobs?"

The family's comments illustrated to me the heterogeneity of environmental attitudes in Futian. While agricultural households and others outside the factory compound were acutely aware of the effects of pollution on their health and livelihoods, industrial laborers, investors, and their families consistently downplayed the seriousness of the threat. From a public-health-risk perspective, this was particularly striking to me since these workers, who toiled day in and day out with no masks or eye protection in the billowing smoke, were arguably exposed to the greatest concentration of toxins. As the sociologist Ulrich Beck has noted, "affliction by hazards need not result in awareness of the hazard. . . . Dangers can always be interpreted away" (Beck 1992:75). Like this particular worker, many of the other workers and their family members made reference to their bodies as barometers of health risk, but their discussions with me almost always amounted to a downplaying of the risks they faced.

One way they downplayed the risks posed by industrial pollution was to "other" the pollution problem by asserting that while environmental conditions in the township were not ideal, there were places far worse than Futian. As the zinc-smelter investor, Mr. Zhang, pointed out to me during one of our conversations, local factories were responsible for only a fraction of the local pollution burden since industrialization was rampant throughout the entire region. He said, "It's very bad in Panzhihua because of Pangang [Panzhihua Iron and Steel Company]. Sometimes the whole sky turns yellow with emissions. But it's not bad here. We don't really notice the impact of these local factories." Like his employee, Mr. Zhang was keen to the fact that regulatory officials in Renhe were putting local factories under increased scrutiny. If the Renhe EPB decided to close Futian's factories for noncompliance with emissions standards, the consequences for their families and livelihoods could be disastrous.

Moreover, the factory's role in the regional political economy was increasingly marginal; state-owned enterprises like Pangang remained at once vital parts of the regional economy and symbols of national pride and development. Although the recent fate of many state-owned enterprises throughout China has involved privatization or closure, Pangang has increased its output in recent years and shows little signs of weakening. Small-scale enterprises, by contrast, carried no such visibility or symbolic import; they could be regulated out of existence with virtually no deleterious effects outside of Futian. Industrial workers were also aware of a historical precedent for closing down noncompliant factories in Futian; in 1997, the Renhe EPB closed a small zinc smelter in Tangba, the primarily Shuitian village in Futian's lower valley. In fact, most factory workers and investors lived in constant fear of being shut down by the EPB, and these fears, as I will discuss in the next chapter, turned out to be well founded.

This kind of "othering" of the pollution problem also speaks to the problem of scale inherent in nearly all pollution problems. Part of the perniciousness of air pollution lies in the ease with which it crosses boundaries. With Pangang, the nation's third-largest state-owned smelter, located just fifty kilometers away, air quality in Futian was undoubtedly affected by the regional dispersion of pollutants from an operation of that magnitude. The community, moreover, is encircled by other industrial townships—Longdong, Daxing, and Shilongba—all of which emit air and water pollution from their own factories. Pollution is a regional problem that respects no spatial boundaries.

Given the magnitude of health risks posed by emissions that exceeded World Health Organization guidelines by nearly a factor of ten, and given the increased vigilance of district monitoring officials, who insisted that emissions from Futian's factories were dangerously high, I wondered how certain individuals were able to systematically ignore the tangible risks they faced. Over time, I came to view the industrial workers' and managers' narratives about pollution as "strategic risk repression" since it allowed them to persist in producing industrial products and profits while compromising their own health and the environmental quality of the community and the region at large (Tilt 2006). Even the young family whose livelihoods depended on wages from the zinc smelter freely admitted that their factory produced black, sulfurous smoke; to argue otherwise would be ignore an obvious material fact. They had decided, however, that their livelihood was more important than the risks posed by industrial pollution, in effect practicing a different kind of "environmental oversight."

The workers' strategic risk repression was partly related to their status as migrants. In the popular discourse about the floating population in contemporary China, two somewhat different threads have emerged. On the one hand, migrant workers are seen as rootless people with low cultural

FIGURE 5.5. A migrant worker in Futian's zinc smelter.

quality, a horde of moving bodies eating the scraps from China's economic table. On the other hand, floaters are sometimes valorized as risk takers, entrepreneurs bravely striking out on their own amidst the dynamism and uncertainty of the reform era (Ngai 2005; Zhang 2001). Many of the migrant workers in Futian described themselves, with a measure of pride, as being capable of "eating bitterness [*chi ku*]," by which they meant that they were willing to travel far from home, endure harsh living conditions, and even expose themselves to toxic contamination if it meant creating economic opportunities for themselves and their families. When describing their work schedules, a number of the young factory workers commented to me with bravado that "we don't have any days of the week here [*zheli meidei xingqi*]," meaning that they frequently toiled seven days per week without a break, as long as there was work to do. A typical work shift began at four A.M. and lasted twelve hours. The free-wheeling atmosphere of the reform era contributed to this willingness to eat bitterness; as migrant laborers, factory workers lacked access to local agricultural land and thus had no economic safety net. Their only chance for success in China's highly unpredictable economy depended on just this sort of self-sacrificing behavior.

REFRAMING "THE ENVIRONMENT" AND "ENVIRONMENTAL VALUES"

Contrary to what might be predicted by much of the literature on environmental values in global context, villagers in Futian were generally aware of and concerned about the air and water pollution caused by local factories. Despite the central role played by rural industry in community development, most people recognized the need for environmentally responsible growth. In fact, growing public awareness of the pollution problem in Futian soon began to motivate government officials to enforce existing environmental laws and regulations by closing local factories, an issue I take up in the next chapter.

Nevertheless, local views about pollution were far from homogeneous. Villagers who relied on agriculture for a living experienced daily the effects of pollution on their health, on their livelihoods, and on the integrity of the agro-ecosystem. Industrial laborers and investors, meanwhile, practiced strategic risk repression, ignoring or downplaying the threats posed by pollution within the calculus of outsized remuneration from factory jobs. This serves to illustrate an important point: environmental values are never politically neutral but are embedded within specific sociopolitical and economic structures. As Ulrich Beck has written, "the same pollutants can have quite different meanings for different people, according to age, gender, eating habits, type of work, information, education and so on" (Beck 1992:26). Farmers, as well as the workers and managers inside the factory walls, tended to situate their attitudes about pollution in the context of key social and economic processes taking place in the community, including liberal economic reforms, factory privatization, an influx of migrant laborers from outside the community, and an increased scrutiny of local factories by environmental regulators.

In fact, these same processes were responsible for the uneven distribution of environmental damages from industrial pollution. Factory privatization led to uneven access to and unequal benefits from local industry. Whereas factories had been viewed as "communal capital" during the collective period of high socialism, in the reform period they became the property of outside investors with little interest in the long-term future of Futian. Farmers, who because of their marginal economic position were perhaps most vulnerable to environmental degradation, bore the brunt of the environmental and economic damage caused by industrial pollution. Their marginal position within the local political economy was mirrored by a heightened awareness of environmental problems. To a certain extent, environmental issues constituted a moral concern for many villagers since environmental degradation was linked with the radical upending of socialist ethics and a new emphasis on individual rather than collective benefit during the reform era. Factory privatization created a group of privileged "outsiders" for whom industry provided generous, if uncertain, compensation. When industrial workers did recognize pollution as a problem, it was seen as a necessary by-product of an imminently desirable goal: economic development.

In short, when we study environmental values, we cannot decontextualize them from other components of social, cultural and economic life. The anthropologist Robert Weller, in a recent book on the influx of global environmental values into China and Taiwan, points to a cultural slippage that may help us interpret environmental consciousness and values in the Futian case. In one of his studies, conducted in rural Anqing in the central Chinese province of Anhui, villagers exhibited extremely low levels of environmental awareness, even after the widely publicized campaign to close down heavily polluting paper mills on the Huai River in the northern part of the province, which had caused major water pollution following floods in the upper part of the watershed (Weller 2006:141).

Because these rural people were not particularly well-off, one might see this apparent ignorance as a confirmation of the postmaterialism thesis. But it is just as likely an example of a disjuncture between an abstract, Western notion of "environmental consciousness," which has been shaped by global elites and discourses that take the human-nature dichotomy as a given, and local understandings of the environment in China, which are grounded in both cultural norms (such as folk religion) as well as a pragmatic brand of environmentalism born of sustained contact with the land. As Weller points out, his informants "lack environmental consciousness only in the sense that they are not concerned with the same issues as national and global elites, or as people who write questionnaires about values" (Weller 2006:157). This raises a fundamental question: Do terms such as "environment" and "nature" call to mind similar images for people in Boston and Beijing, in Chicago and Chengdu?

Much of the Western literature on environmental values is shaped by a biocentric perspective born of urbanization and a consequent disarticulation between people and the natural world. Only when nature becomes something "out there" can it be imagined as a pristine object in need of human protection. But these abstract notions were far less salient frames of reference in Futian than the pragmatic, long-term view of the environment most villagers espoused as a result of intimate contact with and economic dependence upon the land. The ethic of "eating from the mountain" encouraged villagers to reflect on how human activities shaped the local ecosystem. Their ideas about the environment had little in common with an abstract notion of "nature" or "wilderness" devoid of human influence. On the contrary, villagers' understandings of the environment, which were inextricably linked with their economic livelihoods, had a fundamentally anthropocentric component to them.[6]

Cross-cultural research on environmental values often encounters the challenge of capturing salient ideas about what exactly constitutes "the environment" in specific cultural contexts. For example, the traditional semantic categories in Chinese for "nature" were "*tian*" (literally "heaven") and "*shanshui*" (literally "mountains and water"). It was only in the twentieth century that China imported such terms as "nature [*ziran*]" and "environment [*huanjing*]" from the Western ecological sciences via Japan, which had opened its doors to the West decades earlier (Weller 2006:43). I found this to be problematic in my own standardized surveys and interviews in Futian, which began with broad questions about informants' attitudes toward "the environment [*huanjing*]" before working up to specific questions about local factories and pollution. After several extremely tedious interviews in which puzzled informants had difficulty answering my questions, Ms. Duan Shoufeng, a college-educated woman in the Office of Agricultural Technology, informed me as gently as possible that the problem had less to do with my pronunciation and more to do with the semantics of "the environment" in contemporary Chinese usage. The term comprises two characters that, taken together, mean "that which surrounds." But the term is almost always used in tandem with other words that give it contextual meaning; one can speak of the "economic environment [*jingji huanjing*]," the "social environment [*shehui huanjing*]," and even the "political environment [*zhengzhen huanjing*]." In other words, the environment was not an unmarked category; in order to communicate my intended meaning, I had to learn to speak with villagers about the "ecological environment [*shengtai huanjing*]."

6

CIVIL SOCIETY AND THE POLITICS
OF POLLUTION ENFORCEMENT

This [economic] miracle will end soon because the environment can no longer keep pace. . . . We recently shut down 30 projects, including several power plants, [because] the companies involved failed—as required by law—to review what effect their new investments would have on the environment. . . . My agency has always gone against the grain. In the process, there have always been conflicts with the powerful lobbyist groups and strong local governments. But the people, the media, and science are behind us. In fact, the pressure is a motivator for me. Nobody is going to push me off my current course.

—PAN YUE, PRC DEPUTY MINISTER OF THE ENVIRONMENT

IN FUTIAN, the agrarian livelihood was the primary frame through which villagers conceived of and expressed their environmental values. Consequently, factory emissions were perceived as a threat to the agro-ecosystem and to the subsistence activities supported by the ecosystem. I wish to build upon this concept in the present chapter by showing how villagers' actions and responses to pollution, mostly motivated by concern for their livelihoods, contributed to enforcement of pollution standards and ultimately to the closure of local factories at the hands of Environmental Protection Bureau officials. As the key agency representing the Chinese state in matters of environmental quality, what role does SEPA play in establishing and enforcing pollution standards? From where do agency officials derive their mandate to act? Such questions relate to the growing institutional capacity of regulatory agencies such as SEPA, but they also have to do with the burgeoning of civil-society organizations and their intersection with environmental issues in contemporary China.

MOBILIZING THE NATION
FOR POLLUTION CONTROL

China's seemingly intractable environmental problems, of which rural industrial pollution is but one, have led to the perception in the West that its
government is unconcerned about the environment. In the classes I teach
on contemporary Chinese society and on environmental anthropology,
students tend to think of the Chinese environment as a kind of regulatory
"wild west" where anything goes, particularly since lax environmental laws
seem to be one of China's comparative advantages in the global economy,
attracting dirty industries funded by foreign capital.

This perception, while not without some basis, is overly simplistic. Despite its obvious and persistent pollution problems, China is currently
deepening its institutional and fiscal commitments to environmental protection. For example, during the Ninth National People's Congress in 1998,
amidst massive cuts in the national bureaucracy, the State Environmental
Protection Administration not only survived, but was granted more administrative authority (Jahiel 1998).[1] In March 2008, the National People's Congress voted to elevate the agency to full ministerial status, renaming it the
Ministry of Environmental Protection. This originally small environmental-protection bureaucracy, begun in the late 1970s with a few thousand
employees nationwide, has grown into a powerful institution with a broad
mandate to draft environmental laws, conduct environmental-impact as-
sessments, and monitor and enforce nationally set emissions standards (the
administrative structure of SEPA is shown in figure 6.1). In its recent Five-
Year Plan for Environmental Protection (2006–2010), SEPA set ambitious
goals for cutting emissions of key air and water pollutants. Chinese pollution standards are similar to, if slightly more lax than, those of developed
countries such as the United States.

Although its power within the central administration remains limited in
comparison with other state agencies whose mandate rests firmly in the
arena of economic development, such as the State Development Planning
Commission and the Ministry of Construction, SEPA is unarguably gaining
capacity and momentum. In January 2005, for example, Chinese citizens
and the world watched as the agency temporarily halted thirty large-scale
industrial projects, many of them with considerable investment from municipal and provincial governments, for failure to conduct proper environmental-impact analyses. Several of these projects were massive hydroelectric facilities that had been in the works for years and were collectively

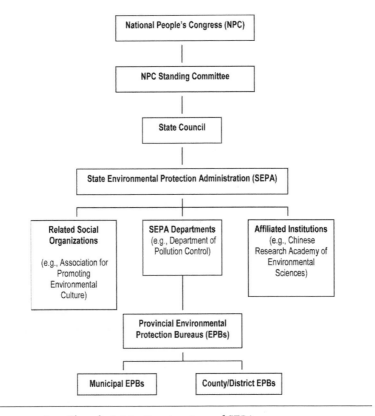

FIGURE 6.1. The administrative structure of SEPA.

valued at billions of dollars. As the words of Pan Yue, deputy minister of the environment, suggest, SEPA's mandate runs "against the grain" of China's headlong rush to industrial development. This bold move was one of the first serious tests of the national Environmental Impact Assessment Law, which was passed in 2003, and a legitimation of SEPA's growing authority (*Renminwang* 2005).

China's fiscal expenditures on environmental protection also underscore its commitment to the cause. Nationwide, these expenditures have risen steadily over the past thirty years, from a tiny fraction of GDP in the 1980s to 1.4 percent of GDP in 2005 (Zhang 2005). While this is slightly less than the amount the United States spends on environmental compliance, which is estimated to be about 2.6 percent of GDP, and less than most other developed countries, it is similar to many developing and middle-income countries. Because China's economy technically still functions under a central plan, the National Development and Reform Commission releases five-year plans for economic development. The current plan, which spans 2006 through 2010, calls for significant further increases in national investment in environmental protection. As I explored government statistics specifically related to pollution control and treatment, I discovered that government subsidies to promote environmental monitoring and mitigation at factories of all sizes grew from 8.2 billion yuan in 1999 to 16.3 billion yuan in 2003 (Chinese Statistical Bureau 2004).

Rural factories have attracted a great deal of attention from the environmental-oversight bureaucracy since the 1990s. Chinese scientists and policymakers acknowledge a wide array of systemic environmental problems with rural industry, including a high rate of natural-resource consumption, a lack of investment in pollution-control technology, and largely unquantified effects on human health and ecological systems in rural areas (Ren and Li 2002; Meng 1999). A number of legislative and policy tools provide SEPA with a mandate to enforce pollution standards in the rural industrial sector. The bedrock of this policy framework is the Environmental Protection Law, which was implemented on a trial basis in 1979 and formalized by the Standing Committee of the National People's Congress in 1989. Chapter 3, article 16 of this legislation gives local governments throughout the country a great deal of power and discretion over protecting the environment. According to article 16, "Local people's governments at various levels shall be responsible for the environmental quality of areas under their jurisdiction and shall take measures to improve the quality of the environment."

This local autonomy, codified in China's foundational Environmental Protection Law, is known as the environmental-protection target-responsibility system (*huanjing baohu mubiao zerenzhi*). In a sense, it represents an application of the Reform and Opening logic to environmental issues as the central government continues to retreat from many of its heavy-handed oversight duties. This delegation of control to the periphery has an important practical implication: the responsibility of overseeing environmental compliance firmly rests with the Environmental Protection Bureaus, which are located within municipal and county-level governmental units. Numbering about 2,500 nationwide, the EPBs represent the frontline defense against pollution and the extension of SEPA's authority to the most peripheral areas of the country.

China's pollution-levy system is the primary instrument through which EPB officials exact environmental compliance from factories. Article 18 of the Environmental Protection Law states that "in cases where the discharge of pollutants exceeds the limit set by the state, a compensation fee shall be charged according to the quantities and concentration of the pollution released." By 1996, this comprehensive levy system, overseen primarily by local EPBs, was in effect in nearly all counties and cities. The levy system is based on self-reporting by polluting factories, which are required to register with their respective EPBs and to disclose information regarding their financial outlook, production-process diagrams, noise pollution, and the so-called three emissions (*san fei*): air pollution, water pollution, and solid waste (Wang and Wheeler 2005).

In practice, however, this delegation of control has had some unintended consequences. EPB officials exercise considerable discretion over how to formally identify factories as noncompliant, how to prioritize their monitoring and enforcement efforts, and how to exact compliance from factories. This leads to substantial ambiguity in the way environmental enforcement is implemented, which a group of World Bank scholars has aptly dubbed the "political economy of regulation" (Wang et al. 2003; Dasgupta, Wang, and Wheeler 1997). In short, EPB officials must weigh the ecological and health consequences of industrial pollution against the economic and fiscal benefits of industrial production. Their jobs are complicated by the fact that EPB officials possess limited organizational resources and thus must prioritize their enforcement efforts.

THE ANATOMY OF ENFORCEMENT

Conducting fieldwork in Futian gave me the rare opportunity to observe China's environmental-oversight process as it unfolded on the ground. The process was at times quite convoluted and difficult to understand, but in retrospect I have come to realize that it was driven by a complex array of political, economic, and cultural factors. Although I was able to witness key proximate events in the enforcement process, other events preceded my arrival in the township. I have thus pieced together a portrait of the environmental-enforcement process by interviewing officials, analyzing published documents, conducting retrospective interviews with factory owners, and conducting participant observation in the township.

In 1996, the State Council, China's primary legislative body, issued *Decisions Concerning Certain Environmental Protection Issues* (Chinese State Council 1996). One section of this legislation was devoted to addressing the growing problem of rural industrial pollution, singling out for closure fifteen types of township and village enterprises that were considered particularly deleterious to the environment. These "fifteen smalls [*shiwu xiao*]" included notorious polluters such as pulp and paper mills, tanneries, and zinc smelters. SEPA scientists had known for years about the harmful effects of water pollution, especially dioxins and heavy metals, from these types of factories.

The following year, the National Environmental Protection Agency (the predecessor of SEPA), in conjunction with the Ministry of Agriculture and several other national agencies, issued *Regulations Concerning Environmental Protection at Township and Village Enterprises*, which gave EPBs a stronger mandate to enforce emissions standards in rural areas. The World Bank also published *China's Environment in the New Century: Clear Water, Blue Skies*, a damning report on the state of China's environment, which closely examined the rural industrial sector as a primary culprit (World Bank 1997). Early in 2000, Xie Zhenhua, then serving as minister of SEPA, spoke at a national conference on environmental protection, where he emphasized that the government would continue to pursue and prosecute small-scale factories that failed to meet emissions standards. The *China Daily* reported on this news with an article that ran under the headline "Headway in Curbing China's Worsening Environmental Pollution" (*China Daily* 2000). This new regulatory framework succeeded in closing some of the worst polluters nationwide, including more than 4,000 small-scale factories in Sichuan province alone (China Environmental Yearbook 2001). Their effect in Futian was negligible, however, where the newly privatized factories continued their operations unabated.

The Renhe District EPB oversees environmental compliance in Futian. It had 14 townships and more than 120 industrial factories under its jurisdiction and conducted regular inspections of the factories under its charge, although limited funding and manpower prevent it from doing so more than every other year. In 1999 a group of inspectors conducted air quality monitoring in Futian, the results of which revealed that sulfur dioxide emissions at Futian's zinc smelter were in violation of SEPA standards. Using the pollution levy system as its guide, the EPB issued a fine and ordered Mr. Zhang, the smelter's main investor, to purchase a smokestack that would vent emissions higher into the air, thereby mitigating some of the health effects for workers and residents in the immediate vicinity. Mr. Zhang and his wife and son complied by spending 40,000 yuan on the smokestack; a portion of this investment came from the factory's discharge fees, which EPB officials returned to Mr. Zhang in accordance with the pollution levy system (Wang et al. 2003:247).

But the problems for Futian's factories were far from over. In 2000, several members of one production cooperative in Wuzitian village, angered and emboldened by the fifty-yuan compensation package offered to them by the district EPB in exchange for the sullying of their irrigation water by the coal-washing plant, which continued to pump out untreated effluents, formed a coalition. Together, they lodged complaints with the township government and with the district EPB. Health concerns were only secondary; the farmers' primary complaint was that the pollution threatened their livelihoods since contaminated water could not be used for irrigation or for watering stock animals. Even after the settlement, members of the cooperative, now feeling slighted by factory managers, township cadres, and EPB officials, continued to complain to the EPB.

Meanwhile, the national legal framework for enforcing air-quality standards in rural areas grew more stringent. In September 2002 the State Council issued a new directive to "strictly implement acid rain control in the two control areas," one of which includes Sichuan, where coal reserves are known to be particularly high in sulfur content. Designed to curb the growing acid rain problem, this plan required each prefecture and city within the two control zones to design plans to meet national emissions standards for sulfur dioxide by 2010. Futian's zinc smelter, which technically should have ceased operations six years previously under the "fifteen smalls" directive, was now the last in operation within Renhe District, probably because Futian is the westernmost township in the district and any monitoring or oversight by EPB officials required a fifty-kilometer Jeep ride over precarious roads.

The watershed event in the enforcement process occurred in late December 2002. A news crew from Sichuan Television Station, which is operated by the provincial government, was alerted by the coalition of local farmers and arrived in Futian to film an exposé program on pollution enforcement for a soft-news television segment called *Jinwan Shifen* (Ten Minutes Tonight). Township cadres were very agitated about the fact that Futian had been under the bright lights of the media. They waited with considerable anticipation for several days before the program aired. On the evening of the broadcast, my wife and I happened to be chatting with one of the owners of a small retail shop on the main road in the open district. The television was tuned to the Sichuan station, and the three of us shelled and ate peanuts while distractedly watching a drama set in the Qing Dynasty.

When the program was over, we watched in amazement as Futian appeared on the television screen. A reporter, wielding a grainy and wobbling hidden camera, interviewed Mr. Zhang, the owner of the zinc smelter, who, oblivious to the fact that he was being filmed, explained how his factory was immune to regulation because it constituted a vital source of tax revenue for the local government. He wasn't worried about environmental regulators, he maintained, because his smelter was so lucrative, not just to him personally but also to the township. The reporter also interviewed several workers from the smelter, and the broadcast interspersed their comments with wide-angle shots of the factory belching black smoke into the sky. Mayor Zhang and Party Secretary Wang also appeared on the program, discussing the state of the township's industrial development and its impact on the local environment and the health of the township's citizens. No one could have predicted it at the time, but this media scrutiny had immediate and lasting consequences. In January 2003, within a few weeks of the television broadcast, Renhe District EPB officials conducted an unannounced inspection and issued a written order that all factories in the township must close because of noncompliance with emissions standards pending further notice. Copies of the order were distributed to factory investors and township cadres.

The story of the zinc smelter and its closure has come to epitomize for me the financial losses that were at stake for local investors and laborers and for the township government. Mr. Zhang, its primary investor, had relocated along with his family from Guizhou in the late 1990s after retiring as a secondary-school teacher. He had formed a business partnership with his wife and adult son, and the group had invested Mr. Zhang's retirement savings, along with money from personal bank loans (approximately 450,000 yuan in total), in the purchase, enlargement, and improvement of the smelter, which had been under the collective ownership of the township government since its construction a decade earlier. Twenty men, most of them from Guizhou, worked in the smelter, and many had brought their wives and children to live with them in the industrial compound. The prospect of permanently closing the smelter was deeply troubling to the investors, who stood to lose most of their capital investment, and to the workers, who would lose their jobs.

Because the zinc smelter's investors were vocal in their opposition to the factory closures, their factory became the site of some of the bitterest controversy surrounding the EPB-mandated shut-down in Futian. Furious about the EPB director's decision, Mr. Zhang told me, "I want him to look me in the face when he tells us to shut down." He soon got his wish: immediately following the Spring Festival holiday (chunjie) in early 2003, which celebrated the advent of the year of the sheep, the monitoring-station chief at the Renhe District EPB agreed to come to Futian to meet with the investors and discuss the fate of their factories. Accompanied by three bureaucrats from the agency, the monitoring-station chief ascended the scrub-covered hill leading to the factory compound in a Jeep bearing the EPB logo, parking in front of the row of dilapidated workers' houses adjacent to the factory. Two of the smelter's six furnaces were in operation, both producing steady black puffs of noxious smoke. The air smelled vaguely metallic.

The monitoring-station chief had visited Futian immediately after the television program had aired to inform the township's factory owners of the EPB's demand that local factories close; it was his intention during this visit to ensure that all factories were in compliance. After exchanging pleasantries and graciously accepting a glass of orange juice from Mr. Zhang's wife, the monitoring-station chief sat down next to one of the furnaces and began the business of negotiation with Mr. Zhang, which I was able to observe firsthand:

> MR. ZHANG: Did you get the report I sent to your office about our business practices and air-pollution emissions?
>
> MONITORING-STATION CHIEF: I got it. But I've told you before, there's nothing I can do. Your factory doesn't meet emissions standards. The smoke is even worse than it was the last time I came up here. It's terribly dirty.
>
> MR. ZHANG: There's some smoke, but it's very clean. It's not harmful. Anyway, you have to give us more time. We are retired, and we've invested our savings in this [factory]. You have to just give us three more weeks to finish off the zinc ore we've purchased.
>
> MONITORING-STATION CHIEF: I've given you enough time. We agreed last time I came here to monitor that you'd close the factory. It's been several weeks now.
>
> MR. ZHANG: [getting agitated] But we did close. We've only got two furnaces out of six running now. Look over here, and you tell me: are the other four furnaces hot or cold?
>
> MONITORING-STATION CHIEF: Okay, okay. Calm down.
>
> MR. ZHANG: Feel them. Are they hot or cold?
>
> MONITORING-STATION CHIEF: [placing a hand on one furnace] They're cold.

At the end of this exchange, Mr. Zhang and his wife presented a series of requests to the district EPB. First, they requested to keep the factory open on a provisional basis for three more months in order to repay as much of their loan as possible and to use up the remaining raw coal that had already been purchased to fuel the smelter furnaces. Second, they requested that the Renhe District government formally recognize in writing that the investors had put their personal money, along with loans, into the enterprise and

FIGURE 6.2. Remnants of the zinc smelter after its closure.

that, as a result, the smelter was to be considered private property. Finally, the investors requested that the district government take steps to mitigate their financial damages. In the end, the investor's pleas for compensation went unheeded by district officials: the zinc smelter, along with Futian's coking plant and coal-washing plant, was forced to cease operations completely. Investors recouped only part of their losses by selling off factory equipment and leftover stocks of raw coal.

For Mr. Zhang, the EPB enforcement activities were ultimately about the rule of law in contemporary China. In a foul mood several days after the visit from EPB authorities, he told me that China's main problem was that rather than being governed by the rule of law (*fa da yu quan*), it was a country in which power and influence routinely trumped the law (*quan da yu fa*). We were sitting on wooden stools in front of his house, overlooking the smelter, which lay dormant, and the main valley of Wuzitian village below. Many of the farm plots were planted in a winter crop of peas, their sprawling vines covered in tiny white flowers. Mr. Zhang distractedly fed a bowl of rice and table scraps to his family's guard dog, an emaciated creature on a chain who never quite warmed to my presence, growling at me every time I came near. To illustrate his point, he showed me a series of glossy brochures, published by the district government under the title *Renhe District Investment Guide*, which were designed to attract outside investors to the area. The documents cited the rich natural resources in the region, including coal, zinc, titanium, vanadium, and iron ores, and outlined plans for developing local industry by offering one-year tax holidays and three-year periods of reduced taxes for new investors. What was an entrepreneur to do, Mr. Zhang wondered, when the government sent mixed messages by attracting private capital to the area and then summarily forcing them to close, without financial compensation, because of environmental infractions? "Our investment is great, and so is our risk [*touzi da, weixian ye da*]," he said.

In my conversation with Mr. Zhang, I couldn't help reflecting on the possibility that my presence as a foreign researcher in Futian helped to precipitate the media coverage of his factory's environmental infractions. It was difficult to ignore the synchronicity between my arrival and the increased scrutiny of environmental regulators. My research plans and study design were on file with the Office of Foreign Affairs at Sichuan University and with the many local political offices in Panzhihua who had granted me permission to conduct the study, so provincial and local officials clearly knew of my ongoing work in Futian. Mr. Zhang assured me that these recent events were not unexpected but were in fact part of a larger environmental-enforcement trend that he and other factory owners had been watching closely for some time. Several months later, when I had the opportunity to ask members of the Sichuan Television Station crew who had covered the Futian story about whether their reporting was in some way related to the probing of a Western anthropologist, they were emphatic that it was not. I will likely never know for sure.

As I pieced together the pollution enforcement process in Futian, I came to realize that it was driven by a range of factors (shown in figure 6.3). SEPA policy served as an ultimate driver of regulatory behavior since it provided the policy structure within which the district EPB had to operate. However, my interviews with EPB officials revealed that multiple other factors were weighed in the decision about how to prioritize and carry out their enforcement duties. Economic and fiscal concerns were of particular importance. Officials had to judge the financial solvency of a given factory based on the factory's own reports; firms facing an adverse financial situation were, by law, less likely to incur strict enforcement (Wang et al. 2003). If the factory constituted a vital revenue source for either the township or district government, it was likewise less likely to be the target of strict enforcement. Although the factories in Futian were a significant revenue source for the township government, their effect on the economy at the district level was negligible since the district contained some 120 factories, most of which were larger and more technologically advanced than Futian's.

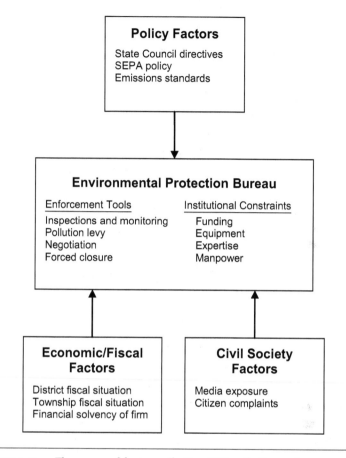

FIGURE 6.3. The range of factors affecting EPB pollution enforcement.

I found it particularly interesting that citizens' complaints and media exposure, what might broadly be called "civil-society factors," provided the immediate impetus for EPB officials to take decisive and strict regulatory action (see Tilt 2007). The township's factories, particularly the zinc smelter, were known to be in violation of national emissions standards for years while receiving only minimal sanction, mostly in the form of pollution-discharge fees. The media played a key role in this case; the provincewide television program constituted a serious loss of "face," or *mianzi*, for local EPB officials, who appeared at best incompetent and at worst corrupt. In rural China, "face" is possessed by people in positions of authority; it is an essential component in accomplishing official tasks (Yan 1996:133–38). One gains face by being a decisive and charismatic leader; conversely, one may lose face (*diu mianzi*) when one's authority is publicly undercut. Regulatory officials at the EPB admitted that they were deeply shamed by the public airing of their failings on television.

In this sense, EPB officials did not operate in a social vacuum; rather, their regulatory decisions and actions were guided by a range of policy and fiscal factors, as well as by concerns about citizen and media mobilization. These proximate factors were enough to overcome the regulatory inertia posed by counter-factors such as the economic concerns of factory investors and township government officials. Enforcement activities were also certainly hindered by the institutional constraints of the EPB itself, including limited funding, equipment, expertise, and manpower. When I interviewed the monitoring-station chief at the Renhe EPB, who happened to share the surname Zhang with the zinc-smelter investor, he conceded that his entire enforcement team consisted of three technicians and himself and that they frequently felt overwhelmed by their duties, which included monitoring nearly 120 factories spread across 14 townships. Both regulatory officials and township government officials also acknowledged that Futian's remote location probably discouraged regular inspections and monitoring, given the limited staff of the EPB.

When it came to actually enforcing the environmental quality standards set by SEPA, EPB officials had a number of enforcement tools at their disposal, including a regular regime of inspections and emissions monitoring, collecting pollution levies, negotiating with firm managers, and—when these measures failed—the forced closure of factories under their jurisdiction. Officials used increasingly coercive enforcement measures as public scrutiny about factory pollution intensified. After viewing the embarrassing television coverage, the monitoring-station chief told me, "This is a war, and I am on the front line [*zhe shi zhanzheng, wo zai qianxian*]."

CIVIL SOCIETY AND THE ENVIRONMENT

The pollution-enforcement process in Futian illustrates some of the ways in which civil society engages with environmental problems in reform-era China. In using the term "civil society," I am referring to an intermediate realm between the family and the state characterized by collective action around shared values, interests, and goals (Hefner 1998; Seligman 1992). As the CCP has liberalized the nation's economy over the past three decades, it has also gradually reduced the scope of its administrative power, increasing the space within which civil society organizations may operate (Chan 2005; Weller 2005). Yet contemporary China presents us with some special challenges in thinking about civil society in regards to the environment since there is little precedent for the protection of individual rights and the single-party nature of the political system severely constrains collective action. In short, a set of empirical questions arises: Where do civil-society strategies come from? What cultural values, historical precedents, and legal statutes underpin them? In what social and political contexts are they deployed? How do various facets of civil society in China engage with the nation's seemingly intractable environmental problems?[2]

The nature of the civil society I observed in Futian bears little resemblance to what students of Western political history might expect to see. Indeed, there was little of the formal civic institutions, voluntary associations, and open social dialogue that are considered part and parcel of the civil society of the West.[3] This is perhaps not surprising, given the repressive nature of politics in China: from the 1989 crackdown on the student-led democracy movement in Tiananmen Square, to the ongoing suppression and persecution of members of the Falungong religious sect, to the quashing of protests in Tibet during the run-up to the 2008 Summer Olympic Games, China's top political leaders have shown time and again that despite new openness in the economic arena, social order is more important than individual liberty when it comes to politics. The mobilization of civil society in Futian over industrial pollution, hardly an unequivocal environmental victory, reveals a certain caution and circumspection on the part of civil-society actors. Environmental civil society in China is undoubtedly still in its infancy. How, then, are we to analyze its role in contemporary China and judge its successes or failures?

PILLARS AND PROCESSES

While the concept may take a unique form within China's peculiar political, cultural, and economic context, I suggest that civil society comprises two key elements: pillars and processes. Pillars, as the term implies, are sturdy, well established social norms grounded in historical precedent that undergird civil society. Processes, meanwhile, can be thought of as the array of strategies that individuals and institutions use to accomplish their desired ends.

At least two pillars serve to support China's burgeoning environmental civil society. The first, ironically, is a deep historical legacy of public mobilization and protest that is grounded in millennia-old social norms. As early as the twelfth century B.C., rulers of the Zhou Dynasty governed the nation based on the "mandate of heaven [*tianming*]," an idea that served to legitimize their rule so long as they governed virtuously. As a corollary, however, imperial subjects had the right to rebel if their leaders made unjust demands or ruled ineffectively; consequently, the dynastic cycle waxed and waned over the centuries as a given ruling family obtained then lost the mandate of heaven (Starr 1997:52–54). Even during the Cultural Revolution (1966–1976), the height of the socialist experiment and the point at which the power of the Chinese state was nearly absolute, this tradition of public engagement in civic affairs remained viable. Citizens were encouraged to purge intellectuals and other "bad elements" from their midst and also to "struggle" against corrupt leaders who abused their political power. For all its brutality, the Cultural Revolution kept alive this tradition of justified public protest. Civil-society mobilizations today—whether they seek to address political, religious, economic or environmental issues—bear the stamp of this legacy (Perry 2003).

The second pillar of environmental civil society is more formal and institutional in nature. China's recent economic reforms have been remarkable in their speed and scope; political reforms, meanwhile, have been incremental but significant, resulting in a growing framework of codified individual rights and a growing consciousness of these rights on the part of the populace. Two key laws are transforming rural China in this regard: the Administrative Litigation Law (1989), which allows individuals to bring lawsuits against governmental units and officials, and the Organic Law of Villagers' Committees (1998), which established a system of self-government by which villagers elect rural cadres. The effectiveness of such laws in fostering meaningful public participation in the governing process is widely questioned, but official statistics show a tenfold increase in litigation of all types over the past two decades, indicating increased public participation in the legal system (O'Brien and Li 2004; Pei 2003).

As public consciousness grows regarding the rule of law and the contractual obligations that bind citizens to the state, a greater discursive space is beginning to open up, within which the processes of civil society may operate. The political scientists Kevin O'Brien and Lianjiang Li have aptly described this growing phenomenon as "rightful resistance," which they define as follows:

A form of popular contention that operates near the boundary of authorized channels, employs the rhetoric and commitments of the powerful to curb the exercise of power, hinges on locating and exploiting divisions within the state, and relies on mobilizing support from the wider public. In particular, rightful resistance entails the innovative use of laws, policies, and other officially promoted values to defy disloyal political and economic elites.

(2006:2)

It is important to distinguish rightful resistance from open rebellion against existing power structures, which in contemporary China can prove politically ineffectual and downright dangerous to those who are seen as challenging the state. Rightful resistance must remain within the bounds of legality set by the government and, if it is to be successful, must invoke some of the ideas and discourses of the government itself. In this sense, rightful resistance straddles "the border between what is usually considered popular resistance and institutional participation" (O'Brien and Li 2006:2). In Futian, the coalition of farmers who spoke out against factory pollution did so on the basis of several recent SEPA policies and regulations that had been successful in closing small-scale factories throughout the country. In this way, they had the rhetoric of the central government on their side.

These pillars support a variety of processes, or strategies, for mobilizing civil-society organizations around environmental interests. The first is the burgeoning environmental NGO movement. Since the 1980s, the central government has cautiously sanctioned the growth of nongovernmental organizations, which are often referred to as "social organizations," or *shehui tuanti*. These groups address a wide array of social issues, from poverty to health care to environmental protection. Friends of Nature, a group founded in 1994 by Liang Congjie, is widely cited as the country's first environmental NGO. The organization has been quite visibly involved in a number of causes over the years, including biodiversity conservation and the protection of several high-profile wildlife species, including the Tibetan antelope. Many groups have followed in the path of Friends of Nature; the past decade has seen a proliferation of environmental NGOs that have mobilized to combat unsustainable forest-harvesting practices, the loss of indigenous environmental knowledge, the requisitioning of rural lands for urban expansion, and the environmental and public-health consequences of industrial pollution. A recent study identified seven organizational types of environmental NGOs: registered NGOs, nonprofit enterprises, unregistered voluntary groups, Web-based groups, student environmental associations, university research centers, and government-organized NGOs (Yang 2005). Such groups share many characteristics, including strong financial and scientific ties to international NGOs and a reliance on leaders with cultural prestige or political power that serves to buffer the organization somewhat from state interference. In the area of pollution advocacy, one important example is the Center for Legal Assistance to Pollution Victims, which is headquartered at Beijing Legal University. Sponsored in part by a Spanish government agency, CLAPV brings together scholars, lawyers, and scientific experts to support citizens in taking legal action against polluters.

While such groups currently number in the hundreds, they are subject to a dizzying set of controls and regulations from the State Council and Central Committee. These controls include allowing no more than one organization of a type to be registered with the Ministry of Civil Affairs and requiring each social organization to be affiliated with an approved governmental unit (Chan 2005; Yang 2005). Indeed, environmental NGOs occupy shaky political ground, given that the central government is still extremely hesitant to grant complete autonomy to nonstate actors, a discomfort that is embodied nicely by the fact that the Chinese term "nongovernmental organization [*fei zhengfu zuzhi*]" sounds suspiciously like "antigovernmental organization [*fan zhengfu zuzhi*]." As Peter Ho has noted, environmentally oriented citizens' groups lack "both the opportunity and the immediate urgency to openly confront the central government" (Ho 2001:897).

There is often a fine line between formal organizing under the banner of an environmental NGO and the second process I would like to discuss: less-institutional, ad hoc mobilization on the part of concerned citizens. The Futian case makes an interesting point, namely, that ad hoc environmental action at the local level can be quite effective. The coalition of local farmers concerned about the effects of pollution on their livelihoods did not need to "openly confront the central government" because environmental enforcement in China is neither centralized nor uniform. Rather, these concerned villagers, who came from the same production cooperative in Wuzitian Village, focused their efforts on influencing regulatory officials in the Environmental Protection Bureau by lodging complaints and by alerting the provincial media to the environmental infractions of local factories. Similar citizen protests happen every day in rural China. While it is extremely difficult to estimate the frequency with which environmental protests occur, in part because of the difficulty in defining precisely what constitutes a "protest," data published by the central government show a marked increase in citizen complaints to EPBs beginning in the late 1990s and culminating with more than 600,000 per year by the mid-2000s (Mol and Carter 2006).

Environmental Protection Bureaus are the most integral pieces of the pollution-enforcement puzzle in contemporary China. EPB enforcement priorities and actions are guided by State Council directives and SEPA policy, but citizen complaints and media exposure, what I have broadly called "civil-society factors," played a key role in determining the regulatory course of action for officials in the Renhe District EPB. Cultural norms, including the EPB officials' aversion to losing "face," were also crucial in this process. Notably, the political economy of China's rural industrial sector was also an important element in the enforcement process. Factories were formerly held by the township government as collective assets, which meant that any criticism of the factory was ultimately criticism of the state, albeit the local state. When the factories were privatized as part of Reform and Opening, however, industry and state were effectively decoupled, which allowed citizens to complain about industrial pollution without posing any direct affront to state power. By transferring capital and the means of production from the collective into individual hands, the liberal economic reforms are undoubtedly nudging China's political and legal systems toward greater openness.

The farmers of Wuzitian village were not vociferous protestors, marching with signs and shouting demands at local cadres and factory bosses. Indeed, it is not always the case that we can see various, discrete interest groups neatly divided against one another. In their view, villagers were merely asking officials to make good on their professed commitment to protect the health, welfare, and livelihoods of the villagers under their charge by complying with existing laws and regulations. Informal in structure and opportunistic in strategy, the actions of Futian villagers may seem to share common ground with James Scott's (1985) "everyday resisters," but they differ in one important way. While everyday forms of resistance consist of covert acts such as "foot-dragging" designed to cumulatively and subtly undermine state power, rightful resistance works in a more straightforward fashion by demanding adherence to the values, laws, and policies of the state itself. As Anna Tsing has observed, environmental movements, somewhat paradoxically, often gain momentum under tight political regimes. In Indonesia in the 1980s and 1990s, for example, the environment was one of the few topics open to discussion and debate precisely because it was discursively tied to national-development goals and ideals; as a result, "saving the nation and saving the environment became linked causes" (Tsing 2005:217).

Although the actions of villagers in Futian were carefully orchestrated to avoid repercussions, they speak volumes about the political changes taking place in reform-era China. During the Maoist era, the interests of the individual were completely subsumed by those of the state. Inciting one's kin and neighbors to oppose the destruction of their livelihoods at the hands of industrial development would have been unthinkable, first, because the household was not a salient economic unit since production was in the hands of the commune and, second, because the government would likely have shown little tolerance for it. Still, civil-society mobilizations of any kind in China seem to be guided by one axiomatic rule: Do not upset social stability (*shehui wending*), the paramount concern of top political leaders, or there will be consequences.

Finally, the media constitutes a vital avenue for fostering the development of environmental civil society. Given the state's explicit commitment in law and policy to strict environmental standards, environmental issues are increasingly considered nonsensitive by government officials, which means that covering industrial accidents or infractions has become more routine in recent years (Mol and Carter 2006). The vast majority of this media coverage is conducted through legal channels since all major television stations and print media are managed by the government, typically at the central or provincial levels. The Futian case is just one of many examples. To cite another, more high-profile example, when an explosion at the PetroChina plant released more than 100 tons of benzene into the Songhua River upstream from the major northeastern city of Harbin in 2005, local officials attempted to cover up the incident. But a deluge of media reports via newspapers and popular television programs brought the issue into the limelight and ultimately resulted in the firing of Xie Zhenhua, the minister of SEPA, who has the distinction of being the highest-ranking official ever to lose his job over an environmental accident (*New York Times* 2005).[4] More dynamic information technologies such as the Internet will undoubtedly facilitate the public's ability to gain information about environmental issues, although currently only about 5 percent of the Chinese populace—mostly well-off, educated, urban people—have regular access to the Internet (Yang 2005:58–59).[5] All of these processes point to a growing desire for transparency and public participation in environmental decision making.

A common historical saying in Chinese politics holds that "there are policies from above, and countermeasures from below [*shang you zhengce, xia you duice*]." In other words, no matter how solid a grip the central government appears to have on its citizens and on local officials, its control is never absolute; individuals and organizations at the margins always have strategies for getting what they need from the political system. Nevertheless, a cautionary note is in order. Civil-society actors—whether in religious, social, or environmental arenas—occupy a tenuous space in Chinese politics. If the mobilization of civil society in Futian tells us anything, it is that greater openness in environmental decision making will come slowly and incrementally and from working within existing power structures, rather than seeking to overturn them.

7

STRUGGLING FOR SUSTAINABILITY

Over the past few years, there has been more economic development, and more opportunities than ever before. But now our factories are shut down, so we've stopped developing.

—MAYOR HU

A S CHINA'S legal framework for controlling pollution becomes more stringent, pollution control is also becoming a fundamental part of the nation's commitment to sustainable development. Yet cadres in Futian, who faced real social and economic consequences because of the closure of the township's factories, struggled to reconcile their interests, and the fiscal needs of the township, with this national mandate for sustainable development. Recent speeches by high Communist Party officials, coupled with a growing body of environmental law and policy, point to a shift toward a new discourse of environmental sustainability on the part of the central government in Beijing. This discourse borrows heavily from the international discourse of sustainability that gained momentum after the 1987 World Commission on Environment and Development, which introduced the phrase "sustainable development" into the popular lexicon. When viewed in the light of SEPA's growing commitment to pollution enforcement, including the stringent actions taken in Futian, this discursive shift signals the Chinese Communist Party's willingness to acknowledge the importance of environmental sustainability in its development goals.

As the Futian case illustrates, the rural industrial sector represents one of the most salient environmental and health risks currently faced by China's population. Because of its prominent role as the engine of economic growth in the countryside, it also represents a focal point in the growing controversy over sustainable development and the necessity of balancing environmental protection and economic growth. Taking a closer look at the consequences of pollution enforcement in Futian allows us to see how the national discourse of sustainable development operated at the local level, how SEPA officials negotiated their job duties within this discourse, and how the enforcement activities of SEPA officials affected the social and economic life of the community.

THE AFTERMATH OF POLLUTION ENFORCEMENT

When it forced the closure of Futian's noncompliant factories, the Renhe District EPB was acting well within its authority since SEPA gives it a broad mandate to protect human health, prevent natural-resource degradation, and avoid economic losses caused by pollution. But EPBs are also embedded within county-level and municipal-level governments, which have an overarching mission to promote economic development and industrialization. By designating "open districts" specially designed for private investment in industry, and by publishing the *Renhe District Investment Guide*, the district government clearly signaled its willingness to nurture the growth of local industry.

Township cadres, who recognized Futian's dependence on industrial revenue, considered it an unfortunate irony that the district government was simultaneously promoting economic development and stepping up its regime of environmental enforcement, which had disastrous economic consequences in Futian. During the course of my field research, cadres allowed me to view the township's financial and budgetary reports. Accompanied by an economist from Sichuan University, I spent several days in an office on the third floor of the government office building reading through these reports, as well as statistical yearbooks published by Panzhihua City and Renhe District.

The documents revealed that Futian's predicament was more dire than I had previously thought (see table 7.1). In the late 1990s, during what many villagers described as the "golden age [*jinhuang shiqi*]" of township and village enterprises, Futian's factories generated more than 2 million yuan in tax revenue annually. Mr. Zhang's zinc smelter, for example, was assessed a monthly management fee (*guanli fei*) of 400 yuan per furnace, a monthly set tax of 150 yuan per furnace, and an annual land-use fee (*tudi fei*) of 2,000 yuan per furnace (the smelter contains six furnaces).[1] The coking plant and coal-washing plant paid similar taxes and fees to the township government, and these revenue sources constituted the bulk of the township budget. The township's fiscal situation had begun to sour several years before the 2003 factory closures, primarily because of a slowdown in the rural industrial sector throughout China, though local coal supplies had also become severely depleted, which meant that coal had to be purchased from adjacent Huaping County, in Yunnan, and transported by truck across the border. EPB officials had dealt Futian's cadres what seemed like the final, devastating blow; following the factory closures in 2003, industrial tax revenue completely disappeared from the township budget. Essential funds for development projects, road building, school maintenance, and other tasks, were reduced to about one-third of their previous levels.

By 2004 the operating budget began showing signs of a gradual recovery, primarily because the township government received an increased yearly subsidy (*buzhu*) from Renhe District. But the township government continued to stagger beneath a cumulative debt of 17 million yuan, which, according to the Panzhihua Statistical Bureau, amounted to more than nine times the average for townships in the district. Much of this debt, owed to

TABLE 7.1   Selected fiscal indicators for Futian, 1998–2005
(unit = yuan)

Year	Tax generated by rural industry	Total township operating budget	Total township debt
1998	2,710,000	1,580,000	15,780,000
1999	2,160,000	2,400,000	15,680,000
2000	N/A	600,000	14,930,000
2001	1,590,000	900,000	17,240,000
2002	1,020,000	200,000	9,490,000
2003	680,000	200,000	10,140,000
2004	0	450,000	14,000,000
2005	0	540,000	17,000,000

Source: Figures for 1998 through 2003 are from Panzhihua City Statistical Bureau (2004). Figures for 2004 and 2005 are based on an interview with the township Communist Party secretary.

financial institutions and to the Panzhihua City and Renhe District Bureaus of Finance, was attributable to poor returns on investments made in township factories during the period of collective ownership. Although factory privatization brought in some revenue in the form of taxes and fees, it did not change the overall fiscal outlook for the township since investors had balked at assuming the township's long-term industrial debt.

Not surprisingly, industrial development remained the main priority of township cadres, whose reactions to the factory closures were somewhat fatalistic. Little Hu, the official in the township Office of Industrial Development, told me with a shrug that nothing could be done about the factory closures: "The upper level of government has spoken [*shangji zhengfu yijing shuohaole*]." The response from Party Secretary Wang was similarly glum: "One thing's for sure: we're going to be a lot poorer [*women jiu hui geng qiongle*]." Li Hong, the township's vice mayor, noted with a hint of sarcasm that in matters of environmental protection, just as in other areas of policy implementation, the central government creates policy with little regard for its effects on local communities. "The upper level of government invites you to dinner," he said. "But the local government pays the bill [*shangji zhengfu qingke, dangdi zhengfu maidan*]." It was becoming clear to cadres that the privatization of the township's industrial assets was not going to be a panacea for their economic troubles.

For many township cadres, the most vexing thing about the factory closures was the apparent insensitivity of the district government to matters of local economic well-being. Mayor Zhang complained that the district could aid Futian by providing tax incentives for investment in environmental mitigation technologies for local factories—such as coal-scrubbing smokestacks—which would bring emissions down to acceptable levels and allow the factories to remain open. In his critique of the district government, Mayor Zhang invoked the "Three Represents [*sange daibiao*]," a policy put forward by former president Jiang Zemin. Considered a body of "important thought [*zhongyao sixiang*]" by the Central Committee, the Three Represents suggested that the Chinese Communist Party's role in negotiating a positive future for the country lay in doing three things: promoting what the Party felicitously called the "advanced forces of production" (i.e., capitalist development); encouraging the advancement of culture; and representing the basic rights of the "overwhelming majority of the Chinese people." Mayor Zhang complained, "That's the most fundamental part of the Three Represents: to represent the old hundred names [*laobaixing*]. But they're not going to do it."[2]

In fact, under China's pollution-levy system, up to 80 percent of pollution-discharge fees may be remitted back to the offending factory to subsidize investment in environmental mitigation technology (Wang et al. 2003). For factories facing special financial difficulties, pollution-discharge fees may even be "reduced, exempted, or postponed" (Chinese State Council 2003). Both the township government and the zinc-smelter investors petitioned the EPB on this point, arguing that increased pollution-levy fees instead of forced closure should be used as the primary enforcement tool since this would allow the smelter to improve its environmental infrastructure while continuing to provide jobs, income, and revenue to the township. But EPB officials, whose resolve had been hardened by the embarrassing media coverage of their lax enforcement of environmental standards, were unmoved. In a letter written to the Panzhihua Bureau of Public Finance, township cadres underscored the seriousness of the situation. Striking a tone of desperation, they pleaded with officials to forgive some of the township's debt, ending the letter by stating that "there is no guarantee that [Futian's] cadres will even have food to eat, let alone be able to pay back the debt."

DEVELOPMENT IN REVERSE

Increased environmental enforcement forced township cadres to consider alterative strategies for economic development. At least for the short term, the township returned to a reliance on agricultural production as its primary economic activity. Since industrial tax revenues dried up, township cadres began trying to develop an agricultural economy of scale—focused on increased production of cash crops such as mangoes and melons—by coordinating between the township's twenty-one small production cooperatives, which are vestiges of the socialist economy, and distributing the township's agricultural products to markets in Panzhihua. In an economic-strategy document that was distributed to the heads of each production cooperative, cadres encouraged villagers to plant honeydew melons, with the goal of increasing the acreage of this crop from 400 *mu* to 1,500 *mu* by 2010.

Farmers throughout China must continually cope with shifting conditions within an economy that is in transition from state-controlled to market-oriented. On the one hand, many of the strategies recommended by township cadres—pooling labor during the busy planting and harvesting seasons or using the production cooperatives as marketing and distribution centers, for example—recall the collective work teams of the socialist period. On the other hand, most farmers participate in such practices out of economic necessity: they need any increase in cash income since the central and provincial governments continue to implement Reform and Opening by pulling back from the provision of key services, such as education. Because grain prices remain under the control of the central government, economically savvy farmers in Futian chose to forego planting rice and sow as much of their land as possible with more lucrative cash crops, gambling that they would bring in enough cash to buy the things they needed.

Of the more than one hundred industrial workers employed in local factories, most were migrant laborers; many opted to migrate elsewhere in Sichuan in search of similar work in other factories, while some have been given local *hukou* status and small land plots by the township government. A few migrant families from Yanbian District, in northern Panzhihua municipality, decided not to return home, setting up residence instead on the hillside just below the factory compound. They built several small shanty houses of wood scraps and plastic tarpaulin sheets, and planted corn, peas, and other row crops on ten *mu* of steeply graded land that the government had classified as "wasteland [*huangshan*]." The soil was dry and rocky, and the families could only irrigate their crops by hand, carrying buckets of water from the stream, which lies several hundred meters down the valley.

Other migrant workers and their families were unsure about what to do. The Yang family, who were still being pursued by the township government over their failure to pay a fee to the family-planning office for the birth of their third child, "the expensive one," were ambivalent. When I asked them about their plans for the future, Mr. Yang's wife replied, "We're just waiting for the last of our salary for this month, then we're going back to Guizhou. But the salary is not what it used to be. The boss doesn't have enough money anymore, so now we only get 300 or 400 yuan per month. But we don't have any choice; jobs are hard to find."

Meanwhile, within a few days of the visit from EPB officials, residents in the open district began commenting on the air quality, which had noticeably improved. The black plumes of smoke which could normally be seen rising over the industrial compound were gone, as was the foul, sulfurous odor in the air. No dump trucks could be seen carrying raw coal to the factory compound, and none carried zinc and coke out. A short climb up the hill to the factory compound revealed a strange setting: factories standing idle and silent; workers squatting in front of their homes, smoking and talking bitterly about the recent turn of events; women distractedly preparing meals; and children, probably oblivious to what was happening around them, playing as usual on the slag heaps. Some of the workers left the factory compound to idly sit in one of the open district's storefronts, playing mahjong or billiards and sipping Jinjiang beer. One young man, a coking-plant worker, jokingly suggested that as a researcher I could measure the vitality of the township's economy in inverse proportion to the number of people playing mahjong on any given day.

The industrial workers and their families felt blindsided by the factory closures. Many took the Renhe District EPB to task for threatening their livelihoods. The wife of Mr. Zhang, the owner of the zinc smelter, made no attempt to conceal her anger at the EPB. She told me, "I know the Environmental Protection Bureau thinks that we pollute. But as far as I can see, there's no harm to people. If you go over to the Renhe District government offices, you tell them that we're dead over here [*women sile*]. People have lost their jobs."

One of the laborers in the smelter, a twenty-five-year-old man from Guizhou, knocked on my door one evening, hastily introduced himself, and begged me to help him find another job. With no savings and no money to get back to his ancestral home, he broke down in tears of desperation. He estimated that he needed a few hundred yuan to pay for a bus ticket into Panzhihua and a hard-seat train ticket to Guiyang, the provincial capital of Guizhou, where he could likely hitch a ride to his village. I told him I would see what I could do, then spent the evening deliberating with Li Jiejie about the best course of action. As we talked, Li Jiejie and I could hear the steady clicking of mahjong tiles being shuffled in the room upstairs, where the party secretary and several other cadres frequently gathered in order to avoid being seen gambling in public.

It struck me that his approaching me was actually a tacit request to simply give him the money. Like most people in the village, he probably assumed that, as a rich foreigner, I could just ask my work unit, or *danwei*, for some cash and give it to him. I was certainly tempted to take this route, but I decided to look for a job for him instead. I spoke with Little Liu, a woman whose family was building a modern two-story house across the street from Li Jiejie's storefront. Her relatives were helping with the labor, so she didn't require his services, but she knew of another construction project in Wuzitian village that needed some skilled laborers for masonry work (which he didn't know how to do) or for coolie work hauling bricks and cement. At fifteen yuan per day, less than half his normal wage at the factory, the pay was not up to his expectations, but he reluctantly took the job. When I asked him how he viewed his prospects for the future, he replied:

It's really hard to say. It's not looking good, though. If we could get the factory started again, things would be better. As it is, even if I can get back to Guizhou, I'd have to find another job, and there's a lot of uncertainty. I don't really want to do the same kind of work. I'd rather find something else. But look at me: I only have a third-grade education. If I had more education, do you think I'd be doing work like this? My cultural quality is really terrible [*wenhua suzhi tai chale*]!

A few weeks after the factory closures, it became obvious that the zinc smelter, coking plant, and coal-washing plant would not reopen. Many villagers turned to scavenging what remained of the industrial materials. One evening, I watched several families climb the piles of rubble with wicker baskets on their backs, digging through the ore slag, rock, and discarded crucibles to find bits of coal to burn at home for heating and cooking. On another day, a work crew dismantled the zinc smelter, sawing apart the metal posts with hacksaws, taking down the smokestack, and hauling away useful scraps of metal and brick. They had paid the factory owner a small fee for the materials. On still another day, a group of villagers loaded scavenged bricks from the smelter onto a flatbed tractor and drove back to Wuzitian village; within a few days, I began to see the bricks appear in the improvised construction of paddy terrace dikes and animal pens. Some ceramic cylinders that had been used as crucibles in the zinc smelter were made into irrigation spouts or dike supports in the rice paddies. Such opportunism on the part of villagers scattered the factory materials around the township and likely provided new pathways for arsenic, lead, and mercury to enter the groundwater, soil, and food system.

The factory closures created unanticipated ripple effects that were felt by nearly every household in the township. For example, agricultural households were normally taxed based on the amount of land under cultivation each season; as of 2003, however, farmers complained that they were now being taxed on their total land holdings, including fields left fallow. A neighbor of Li Jiejie's strolled past the storefront one evening and stopped briefly to rant about "those bastards" in the township government. The woman's family, surnamed Gao, held 1.8 *mu* of good paddy in Wuzitian. Township cadres had demanded that the household pay 133 yuan in taxes for the 2002 fiscal year. She refused, saying that the family had left the field fallow that year in order to focus their energy on running their small retail shop in the open district and so did not harvest any crops. With no agricultural yield, she felt that she was under no obligation to pay her agricultural tax. She appealed to the leader of her production cooperative, who, being new to

FIGURE 7.1. A woman and her daughter work in their rice paddy, with discarded cylindrical crucibles from the zinc smelter piled at the terrace edge.

the job, did not have enough political clout with the government to intervene on her behalf. The woman and her husband had just returned from appealing at the government offices, where they were told that their land would be taxed, fallow or not, since industrial revenue from local factories had completely dried up and cadres needed to stem the financial crisis. Lacking any recourse, the family reluctantly paid.

Beginning in the early 1990s, many villagers had invested their savings in economic-development bonds issued by the township government to fund capital investments in local factories; in effect, the township's dismal fiscal outlook, including millions of yuan in debt, ultimately rested on the shoulders of the villagers themselves. One woman complained to me that officials refused to cash out her development bond, which was worth more than 17,000 yuan, a tactic she viewed as an attempt by the government to shore up the township budget in the face of declining industrial revenue. Finally, the construction of the 2,000-kilowatt-capacity hydroelectric facility on the local stream represented another attempt to find alternative revenue sources. Officials estimated that the electricity, which would be sold to the regional grid, would bring in 1 million yuan per year, although returns have fallen far short of that figure. Cadres also renewed efforts at coal exploration in the township, soliciting capital investment from the

Panzhihua and Renhe governments. This strategy carried considerable risk, however, since the township would have to assume additional debt beyond the 17 million yuan it already owed.

SINICIZING THE DISCOURSE
OF SUSTAINABLE DEVELOPMENT

Underlying the trend toward increased environmental enforcement of rural factories is a conflict of values regarding the precise meaning of "sustainable development" in reform-era China. Chairman Mao Zedong once famously declared that "if people living in nature want to be free, they will have to use natural sciences to understand nature, to overcome nature and to change nature; only then will they obtain freedom from nature" (Mao 1966:44, quoted in Ho 2001:895). Maoist discourse on the environment was heavily influenced by Marxist dialectical materialism, in which the concept of class struggle played a central role. For Mao, the environment was yet another oppressive force that the Chinese people would have to struggle against in their quest for modernity.

In one of his many poems, a pursuit that he ironically continued even amidst the anti-intellectual sentiments of the Cultural Revolution, Mao wrote, "To struggle against the heavens is endless joy; to struggle against the earth is endless joy; to struggle against people is endless joy [*yu tian dou, qile wu qiong; yu di dou, qile wu qiong; yu ren dou, qile wu qiong*]." These lines effectively summarize much of the ideology regarding the environment during the Maoist period: human beings and nature were fundamentally separate, and the goals of humans, which supersede the needs of the biophysical environment, must be accomplished through a reliance on science and technology that effectively "conquers" nature.

At the time he penned these lines, Mao's rhetoric had already been preceded by more than a decade of policy geared toward radically altering the Chinese environment, a pattern that would continue well into the reform period. During the campaign to eradicate the "Four Pests," for example, citizens were encouraged to wage war against rats, flies, mosquitoes, and sparrows, each of which was perceived as a threat to human health or to grain yields. A song frequently sung by commune members in the countryside echoed this sentiment:

> In heaven, there is no Jade Emperor,
> On earth, there is no Dragon King.
> I am the Jade Emperor, I am the Dragon King.

I order the three mountains and five peaks:
"Make way, here I come!"[3]

Still another illustration of the human conquest of nature is reflected in
an ancient story called "Yugong Yi Shan" (The Foolish Old Man Who
Moved the Mountains). Known by nearly all Chinese, the story depicts an
old man who methodically and patiently moved a mountain that blocked
the path to his house, one shovelful at a time. Mao himself invoked this well
known fable in one of his speeches to the National Congress in 1945, during
the Cultural Revolution:

> [The story] tells of an old man who lived in northern China long, long ago
> and was known as the Foolish Old Man of North Mountain. His house
> faced south and beyond his doorway stood the two great peaks, Taihang and
> Wangwu, obstructing the way. With great determination, he led his sons in
> digging up these mountains, hoe in hand. Another greybeard, known as the
> Wise Old Man, saw them and said derisively, "How silly of you to do this!
> It is quite impossible for you few to dig up these two huge mountains." The
> Foolish Old Man replied, "When I die, my sons will carry on; when they
> die, there will be my grandsons, and then their sons and grandsons, and
> so on to infinity. High as they are, the mountains cannot grow any higher
> and with every bit we dig, they will be that much lower. Why can't we clear
> them away?"
>
> (MAO 1971:321)

The story was intended to instill in the peasantry the sense that, given the
proper collective mindset, they could alter the natural world to meet human
needs in the face of overwhelming odds. This was not merely a rhetorical
device: during the early 1960s, when regional self-reliance in grain production was a key policy goal of the central government, members of the Dazhai People's Commune, located in central China's Shanxi province, endured backbreaking labor in order to turn steep mountainous lands of
marginal quality into terraced grain paddies. Photographs of the communal
lands, which often appeared on propaganda posters in a campaign called
"Learning from Dazhai [*nongye xue dazhai*]," depict thousands of terrace
rows carved out of impossibly steep hills. The human effort exerted to con-
quer nature on behalf of a modernizing nation appeared truly heroic.

These early socialist-era ideas about the environment, which depicted
humans and nature as continually locked in an antagonistic battle, are now
being supplanted by new ideas and discourses from abroad. The past
decade in particular has seen a dramatic shift in the attitudes of central-

government authorities toward the environment as the threats to human welfare posed by industrial development become increasingly obvious and as environmental degradation begins to seriously undermine economic growth. The phrase "sustainable development," which became an international catchword after the United Nations World Commission on Environment and Development in 1987, has been adopted and circulated within the Chinese government and within Chinese-language scholarship on the environment.

In state-sponsored publications, sustainable development, most often translated as *"kechixu fazhan,"* or "development that can be sustained," is defined in accordance with the commission (1987:43) as "development that meets the needs of the present without compromising the ability of future generations to meet their own needs [*ji manzu dangdai rende xuyao, you bu dui houdai rende nengli goucheng weihaide fazhan*]" (China Industrial Development Report 2001:96–108).[4] China was a high-profile participant in the 1992 United Nations Conference on the Environment and Development in Rio de Janeiro, Brazil, adopting Agenda 21, a broad policy document outlining environmentally sound development strategies in which the goal of sustainable development was defined as ensuring "socially responsible economic development while protecting the resources base and the environment for future generations" (see Beckerman 1998).

The momentum of official Chinese participation in these international agreements has carried over into domestic policy. Former President Jiang Zemin, for example, in a speech before the Fifteenth National Congress in 1997, cited the size of China's population and the finite capacity of the country's natural resources as key reasons why the country must put forward what he called a "strategy of sustainability [*kechixu fazhan zhanlüe*]" that entailed pursuing economic development while allocating more resources, technology, and scientific knowledge to environmental protection activities (Wang, Wang, and Wang 2000).

Similarly, in his speech before the Sixteenth National People's Congress in 2002, President Jiang discussed the meaning of *"xiaokang"* (literally "small comfort," or being well off) within contemporary Chinese society. Given the ubiquitous use of this term in current policy, the fact that President Jiang made *"xiaokang"* part of his address to the NPC is barely worth noting, but his particular redefinition of the term is. In his speech, Jiang laid out what he called "four main goals in establishing a *xiaokang* society [*jianshe xiaokang shehui si da mubiao*]," the last of which, despite its obtuse language, unmistakably underscores the national government's shift toward a discourse of sustainability:

[China should] continuously strengthen the ability to attain sustainable development [*kechixu fazhan*], better the natural environment, improve efficiency in the use of natural resources, attain harmony between man and nature, push toward the development of the forces of production for all society, a rich life, and the road leading to the development of a better environment and civilized society [*wenming shehui*].

(CHINA TOWNSHIP AND VILLAGE ENTERPRISES 2002: 4)

Similarly, Premier Wen Jiabao, in a 2006 speech at the Sixth National Congress on Environmental Protection sponsored by the State Council, stressed what he called the "three transformations":

1. From an economy-centered model of development to equal attention on both environmental protection and economic development
2. From a mindset of cleaning up after economic development to a simultaneous emphasis on environmental protection *and* economic development
3. From a sole emphasis on administrative measures to control the environment (top-down, hierarchical, institutional) to a more comprehensive system involving the state, the business sector, and civil society

(XUE, SIMONIS, AND DUDEK 2007; CHINESE STATE ENVIRONMENTAL PROTECTION ADMINISTRATION 2006)

Supported by official policy and rhetoric, Chinese scholars of environmental science and policy are also actively exploring the implications of sustainable development for contemporary China. Scholars writing in the new journal *Environmental Protection*, a publication sponsored by SEPA, for example, argued that the pursuit of sustainable development is an urgent issue of national security. Under the banner of official government sponsorship, scientists and policymakers are increasingly advocating a position that requires both the government and citizens to recognize that social-ecological systems must be managed holistically and with the long-term future in mind (Qu 2002). Chinese environmental scientists within SEPA and other agencies are more openly urging a policy shift toward sustainability that would contain five key elements:

1. Working toward intergenerational equity by considering the resource needs of future generations
2. Using the principles of environmental economics in order to achieve sound resource management

3. Balancing the needs of humans and nature
4. Emphasizing environmental education
5. Placing the goal of ecological security on equal footing with economic and national security

(GUAN 2002:16–18)

This shift toward an emphasis on sustainability within government and scholarly circles in China coincides with a gradual burgeoning of the environmental oversight bureaucracy and increasingly strict enforcement tactics, as cadres, workers, and villagers in Futian learned from firsthand experience.

STRUGGLING FOR SUSTAINABILITY

Of course, policy formulation does not always translate to implementation, particularly in China's rural industrial sector, which poses a special challenge to the environmental-protection bureaucracy. Disjuncture between the central government and local governments is a long-standing theme in Chinese political history. At a practical level, enforcement officials within EPBs face a number of challenges as they attempt to regulate rural industry, including ambiguity of laws and policies, overlapping jurisdictions with other agencies, and a lack of technical capacity and expertise (Ferris and Zhang 2005:88–93). Renhe District EPB officials, including the monitoring-station chief, reiterated to me a number of times how difficult it was for their staff members to cover the wide geographical area under their charge. The environmental-oversight bureaucracy is also plagued by some of the same institutional problems as similar agencies in the West; its red tape is so difficult to navigate, and its offices so encumbered by bureaucrats, that it is often viewed by ordinary Chinese people as a case of "ten sheep with nine shepherds [*shi yang jiu mu*]."

The central government is responsible for the creation of environmental policy and the dissemination of environmental protection discourse, but it is the regional- and local-level governments that are left with the job of enforcing compliance, as the factory closures in Futian illustrated. Every county-level jurisdictional unit in China is required to have an Environmental Protection Bureau. While they represent the reach of SEPA into the nation's periphery, these agencies are also typically embedded in county governments, which have an overarching mandate to promote economic development. This often creates serious conflicts between EPBs and other government offices. Tian Weiyong, a former SEPA division chief, acknowledged in a 2002 interview with a journalist, "Some local governments refused to cooperate in punishing those factories [for pollution infractions], which [makes] our work very difficult" (Economy 2004:106).[5]

The Renhe EPB, which ordered the factory closures in Futian, was in fact torn between its dual responsibilities of promoting industrial development and enforcing emissions standards. In the early spring of 2003, not long after the factory closures, I hired a car to take the fifty-kilometer trip to the district town of Renhe to meet with Mr. Zhang, the monitoring-station chief at the EPB. The panzhihua trees were in bloom, their red and orange flowers brilliant among thick branches. I counted more than a hundred of the trees along the way, most of them clustered in valley bottoms near streams and irrigation canals. The blossoms are edible, which explained why the trees were picked clean near more populated areas. The blossoms were also quite expensive, selling in the market for four or five yuan per *jin*, double or triple the price of most vegetables and nearly as expensive as some meats. Served lightly stir-fried with Sichuan hot peppers, they were delicious.

Mr. Zhang, whom I hadn't seen since the day I had observed him negotiating with the owners of the zinc smelter, was an environmental chemist in his early thirties. His demeanor was cordial but serious. In my interview with him, he explained that the most pressing problem facing his agency was sustainable development (he used the same term as President Jiang Zemin, "*kechixu fazhan*"), which he defined as balancing economic development and environmental protection. Reflecting on the role of pollution enforcement in China's new trajectory toward sustainable development, he added:

> It's particularly difficult in backward places, like Futian, that have poor technology. Places with backward technology, heavy polluters, will be phased out. Once you've reached a certain level of development, then you need to start considering environmental protection. This is a difficult thing to do. Even in the West, people destroyed the environment in order to develop the economy.

As a scientist within an EPB located in a rural backwater jurisdiction, the monitoring-station chief's remarks struck me as poignant for their mastery of the discourse of sustainability currently circulating within the highest levels of government in China.

The enforcement of emissions standards was a local act, but the logic underlying enforcement was profoundly shaped by national and international discourses. Although Mr. Zhang knew little about the history of the sustainable-development discourse in the West, he knew that it was a product of policymakers and politicians whose countries enjoyed considerable wealth and that this wealth had been made possible by relatively lax environmental standards during the formative stages of industrialization.

By contrast, township cadres in Futian were less sanguine about sustainable development. Although all of Futian's factories had been privatized by 2002, eliminating industrial profits as a revenue source, the bulk of the township government's operating revenue was still derived from industrial sources in the form of taxes on outputs. The township government's interests thus aligned quite closely with those of industrial investors, and the closure of local factories by the district EPB had disastrous consequences for township finances, as I described above. During the 1990s, Futian relied on industry for nearly 85 percent of its operating revenues, funneling industrial profits into community-development programs. The township government constructed new schools in three of its four villages, paved key sections of the intervillage road, and constructed a new six-story government office building on the edge of the newly created open district. This "Futian Model [*Futian Moshi*]" was held up as a regional example of using industrial development to raise living standards in peripheral areas with heavy concentrations of ethnic minorities, a perennial problem in rural China and the southwest region in particular (Li 1995).

Mr. Tian Lizhong, the government official in the Agricultural Technology Office in Futian, who had also introduced me to many of the intricacies of *minzu* culture, contextualized the debate over sustainability and pollution control by defining sustainability with a colloquial play on words. In one of my many interviews with him, he suggested that "sustainability is trying to solve the *wenbao* problem without exacerbating the *huanbao* problem." "*Wenbao*" (literally, "warm and full") is how many rural Chinese described the standard of living just above poverty, when one's belly is sated and there is a roof beneath which to sleep. It is a standard of living that has been attained in Futian only recently; older residents often related tales of eating grass during especially hard times or slaughtering a single pig to provide the annual meat supply for a family of six. "*Huanbao*" (an abbreviation of "*huanjing baohu*," or "environmental protection"), by contrast, is a relatively new concern for Futian and for much of rural China, and one that has taken on salience only in the last decade or so since rural enterprises have altered the local landscape and industrial pollution has become an intractable problem. The problem of *wenbao* versus *huanbao* ultimately involved a question of values regarding precisely what should be "sustained." Should government officials promote industrial development, thereby increasing the living standards for industrial laborers and investors and providing critical revenue for the township government, or should they take meaningful steps toward preventing further degradation of the local environment, which would also help safeguard agricultural livelihoods?

Table 7.2 provides a characterization of the competing models of sustainable development held by various governmental levels in regards to rural industry and pollution in Futian. In the context of environmental protection and sustainable development, the Chinese government can in fact be viewed as multiple states, each promoting its own model of sustainable development. As many anthropologists have learned, environmental politics often involves "struggles at the ideological and discursive levels including, but not restricted to, definitions of science, knowledge, environment, sustainability, and biodiversity" (Derman and Ferguson 2003:278; see also Tilt and Xiao 2007). These are struggles of meaning that take place in both the economic and ideological dimensions as the central government, district government, and township government debate about the particular model of sustainability that should prevail in regards to rural industry.

The central government, via a growing body of environmental law and policy, promoted a model of sustainability that addressed the *huanbao* problem by implementing and enforcing industrial-pollution-emissions standards. I have come to think of this model as "idealized sustainable development" because it borrows heavily from current international ideals about sustainability and sets environmental-protection policy accordingly but pays little attention to enforcement tactics or their consequences. The "idealized sustainable development" model is hierarchical in structure and modernistic in outlook: environmental problems are to be addressed by policy emanating from the central government and implemented by scientists and bureaucrats armed with technology and expertise.

The township government, on the other hand, put forward a literal, albeit resourceful model of sustainable development, conflating the concept with "sustained development," which it envisioned as economic growth in perpetuity, unhindered by the constraints of environmental regulation. This model of sustainability placed emphasis on solving the *wenbao* problem by preserving the income and tax revenue generated by local industry and using the revenue for much-needed community-development projects that benefit the township's residents. Township government officials viewed industry as the most important path to local economic development,

TABLE 7.2 Characterization of sustainable-development
models promoted by various state levels

State level	Model	Description of objectives
Central government (e.g., SEPA)	Idealized sustainable development	Primary objective is to promote environmentally sustainable development practices through State Council directives, SEPA policy, and pollution-emissions standards
District government (e.g., EPB)	Pragmatic sustainable development	Primary objective is to implement SEPA policies regarding pollution control, but must also consider the fiscal outlooks of the township and district governments and constraints in funding and expertise
Township government	Sustained development	Primary objective is to promote industrial development in order to provide jobs, income and revenue for the township

although for agricultural households, whose livelihoods were undermined by air and water pollution, industrial emissions were considered both environmentally and economically damaging.

Officials in the district government subscribed to a model we might think of as the "pragmatic sustainable development model." They were bound by SEPA policies and directives, but they exercised considerable discretion over how to carry out their enforcement duties. Because they were embedded within the district government, which vigorously promoted industrial development, officials in the EPB had to weigh the benefits of pollution enforcement against the economic and fiscal needs of the district and township. Given the sheer scale of their enforcement duties, they also faced real institutional challenges in the form of manpower, expertise, and funding shortages. In my interviews with them, EPB scientists and policymakers generally projected an image of competence, professionalism, and dedication to environmental protection. But with limited monitoring equipment and a single agency Jeep to aid them in their enforcement duties, the district's factories proved difficult to manage. The pragmatic sustainable development model also reflects the complex role of the state in reform-era China, which is retreating in some areas and simultaneously expanding its power and influence in others.

To a certain extent, the field of international development is predicated on the modernistic assumption that proper governance and the use of technology can turn economic progress and environmental protection into complementary goals. As a former president of the World Bank once put it, "sound ecology is good economics" (Conable 1987: 6). But the competing models of sustainability that arose in Futian pointed to a troublingly ambiguous side of sustainable development, one full of significant, if unstated, contradictions. In a recent essay, for example, two economists reviewed eleven large-scale research projects designed to measure and operationalize the ambiguous concept of sustainable development (Parris and Kates 2003). Many of these projects, including well-known examples such as the Redefining Progress Project and the Genuine Progress Indicator, devise a set of indicators (environmental, economic, social, institutional) and then measure the performance of hundreds of countries based on these indicators. Two interesting trends emerged from this review. The first was the heterogeneity of the indicators themselves: some of the projects were focused rather narrowly on particular issues such as literacy, education, or conservation, while some represented a composite of hundreds of indicators on disparate topics such as hunger, energy use, water use, deforestation, and pollution.

The second interesting trend, and one more germane to the Futian case, has to do with the relationship between "sustaining" and "developing" (see table 7.3). If we separate the two parts of the phrase and ask ourselves, "What is to be sustained?" and, "What is to be developed?" we begin to see some contradictions emerge. We may all agree, for example, that increasing household incomes or educational opportunities are laudable goals for governments and other organizations to pursue, but to the extent that accomplishing these goals depends on an intensification of the extraction and use of natural resources, it becomes a zero-sum equation in which "developing" inevitably takes something away from "sustaining." Indeed, key terms such as "sustainability" often serve merely to highlight the differences between various stakeholders in environmental management problems.[6]

This tension between "sustaining" and "developing" is evident in China's recent development trajectory. Industrial managers and laborers in Futian saw it as the township government's signal responsibility to promote local development in accordance with the promises of the central government.

TABLE 7.3 Balancing between "sustaining" and "developing"

What is to be sustained?	What is to be developed?
Nature	**People**
Earth	Life expectancy
Biodiversity	Mortality and morbidity
Ecosystems	Education
Life-support systems	**Economy**
Ecosystem services	Wealth and income
Natural resources	Productive economic sectors
Environmental quality	Consumption
Community	**Society**
Culture	Institutions
Ethnic and linguistic identity	Social capital
Attachment to place	

Source: Adapted from Parris and Kates 2003.

Often, the rhetoric and policies of the central government were used as tools for leveling critiques against the district government for enacting harsh environmental protection policies. For example, Little Hu criticized EPB officials by citing the Great Western Opening policy (*xibu da kaifa*). Initiated by President Jiang Zemin and built upon by current leaders, the Great Western Opening policy was designed to encourage the influx of investment, both domestic and foreign, into China's poorest interior regions in order to foster economic development and ultimately turn the west into "a new and strong region marked by sound economic development, social progress, welfare, ethnic unity, and with beautiful and clean mountains and streams" (Chinese Statistical Bureau 2001:12).

Great Western Opening was instituted with the goal of remedying the vast economic disparities across regions in China, disparities that were nurtured for two decades by government officials eager to promote development along China's east coast under Deng Xiaoping's proclamation that "some must get rich first." By this standard, Futian's development trajectory in the wake of factory closures was a dismal failure in the eyes of many cadres. Little Hu told me, "The policy [Great Western Opening] is supposed to bring a lot of investment into places like this. There has been a lot of development; workers in these factories made better wages than all the farmers. But now all that's finished." More to the point, for local cadres, the factory closures represented a failure on the part of district government officials to enact key development policies passed down from higher levels of government.

Similarly, for industrial managers and laborers and their families, development was about job creation and enjoying the benefits of the Reform and Opening policies of the last several decades, which facilitated the accumulation of wealth with the goal of establishing a *xiaokang* society. Within this framework, the harsh enforcement tactics employed by Renhe EPB officials seemed incongruous and even counterproductive. Factory managers and laborers thus saw the decline of Futian's industries, and the job loss that such a decline entailed, as a direct attack on their livelihoods by the district government. As one worker in the coking plant pointed out to me:

> This place is supposed to be an open district [*kaifa qu*] for economic development. But how can you develop like this, with no jobs? If they keep shutting down factories, the air will be clean but everyone will be unemployed. Anyway, *xiaokang* is just a word invented by the government. It doesn't mean anything; they just say it because it sounds good [*hao ting*].

It is worthwhile to ask what the central government stands to gain by espousing the international discourse of sustainable development and by enforcing pollution-emissions standards, given that this represents a significant shift from the Maoist and early reform periods. First, even conservative estimates suggest that industrial pollution cuts 4 to 5 percent from the nation's annual GDP in the form of clean-up costs, healthcare costs, and lost wages (Liu and Zhou 2001; see also World Bank 2007). Communist Party officials, whose credibility ironically rests in part on maintaining China's high level of market-driven economic growth, cannot afford to ignore this problem.

Second, by adopting the discourse of sustainability, which is presently enjoying wide international circulation, the Chinese government gains a measure of credibility within the international political arena. A substantial amount of foreign aid and investment flows to all levels of the Chinese government via bilateral aid, green investing, and NGOs because of the central government's effective use of the discourse of sustainability (Vermeer 1998). The World Bank, the United Nations Development Program, and the Asian Development Bank are all major financial backers of sustainable-development projects in China. Securing this investment requires skillful use of the sustainable-development discourse, within which science and technology are seen as the solutions to environmental problems. These problems, moreover, are framed primarily in materialist terms as the biophysical limits of a finite environment, rather than as sociopolitical problems, rooted in frenetic market-driven development, that involve extremely difficult tradeoffs.

Ironically, a major step toward China's participation in the international discourse of sustainability began with the construction of the Three Gorges Dam, the largest hydroelectric facility in the world and a magnet for opposition by environmental groups. The dam, which upon completion will stand 185 meters above the Yangtze and span more than two kilometers in breadth, became a rallying point for international groups concerned about the displacement of more than a million people from the area surrounding the reservoir. For environmental groups, the dam became a symbol of China's modernist commitment to pursue national development at any cost. Environmentalists protested a host of projected environmental problems including the buildup of silt behind the dam, the loss of estuarine habitat, and the destruction of rare species, including the Yangtze freshwater dolphin.[7] Somewhat tellingly, the outpouring of environmentalist sentiment surrounding the Three Gorges Dam has largely been an international phenomenon, bolstered by Chinese intellectuals writing for an international audience. Two books edited by the famous journalist and anti-dam activist Dai Qing in the 1990s, for example, have been banned in China and widely circulated abroad (Dai 1999, 1994).[8] Such international exposure prods the central government toward at least a tacit espousal of environmental sustainability.

At a more pragmatic level, the discourse of sustainability, and its extension through the enforcement of pollution standards, provides an avenue for the central government to exercise power over peripheral areas such as Futian at a time in which the state is retreating from administrative oversight in other ways. Policies designed to protect the environment can often become a way for central states to gain control over other state and nonstate actors (Agrawal 2005). Of course, the central government also has a stake in quelling the popular uprisings over pollution that are occurring throughout China with greater frequency.

The closure of rural factories in hinterland areas such as Futian provides a means for the government to enact environmentally sustainable policies while keeping the economic effects of such policies minimal and localized. Many township cadres and factory investors felt that the EPB carried out its enforcement duties in a selective way that depended ultimately on which factories were capable of defending themselves and garnering the necessary social connections and resources to avoid sanction. For example, despite the fact that emissions from large state-owned enterprises like the nearby Panzhihua Iron and Steel Company surpassed Futian's emissions by an order of magnitude, such state-owned behemoths remained to pollute another day because they were vital sources of employment and revenue, because they carried symbolic and political value for the state, and because they were effectively beyond the regulatory reach of the municipal and district EPBs.

Of course, regulatory officials were well aware of the egregious pollution caused by Pangang; scientists at the Panzhihua Municipal EPB reluctantly gave me a tour of their facilities, which consisted of a highly sophisticated air-quality-monitoring station on the roof of the bureau's office building. The equipment collected real-time data on total suspended particulates, sulfur dioxide, and nitrogen dioxide, feeding the information to a series of computers via an Ethernet connection. There were also instruments to measure wind speed and wind direction, which helped scientists estimate how pollutants were dispersed over the city and the wider region.

In the view of many township cadres, the small-scale factories in Futian bore the uneven consequences of pollution enforcement not because closing them would yield significant cuts in regional emissions but because their closure would result in the least hassle possible while providing concrete evidence of the government's commitment to act on its policies of sustainability. Furthermore, officials in midlevel agencies such as the Renhe District Environmental Protection Bureau lacked the institutional power and jurisdiction to tackle massive, state-owned enterprises like Panzhihua Iron and Steel.

WHAT SHOULD BE SUSTAINED?

As this chapter has shown, the precise meaning of "sustainable development" was a major point of contention as Futian faced a loss of tax revenue, income, and jobs in the name of environmental protection. This raised several interesting points to consider. First, environmental enforcement in Futian underscored the fact that the central government's growing commitment to sustainability, voiced through public discourse, policy, and law, was now increasingly backed by action. The discourse of sustainable development at the central-government level filtered down to the lower-level government agencies responsible for enforcing compliance with pollution standards. In fact, the monitoring-station chief and his colleagues in the Renhe District EPB viewed their enforcement duties as an extension of this national discourse, which had already resulted in the closure of tens of thousands of small-scale factories nationwide.

Second, the closure of Futian's factories encouraged me to move beyond thinking of the Chinese government as a singular, monolithic entity with clearly definable interests. Rather, it became more important for me to examine the divergent positions and interests of various state agencies in regard to the question of sustainable development. The dynamic tension between center and periphery is a common theme in much of Chinese political history, and this tension has, if anything, deepened during the reform period (Whiting 2000; see also Selden 1998). Within the realm of environmental politics, even a tightly controlled single-party state like the People's Republic of China contains contentious positions within different levels of the government, and these positions have important consequences for determining exactly how sustainability is defined and implemented.

Finally, while the roots of intrastate conflict over sustainability were economic in nature, stemming from the continued reliance on industrial revenue at the township and village levels, the conflict was also ideological and value-laden insofar as it hinged on several key questions: *What* should be sustained? For whom? And for how long? The closure of Futian's factories highlighted the contested meanings of sustainable development, with different levels of government advocating wholly different models of sustainability based variously on preserving ecological integrity, promoting community development, and retaining vital industrial revenue. Because the enforcement of emissions standards occurred at the intersection of the environment and the economy, it was an inherently political issue that entailed consequences for the environment and human health, but also for economic growth, employment, and community development. Within this struggle, where key terms like sustainability became the focus of contestation, pollution enforcement opened up deep fissures between the township, district and national governments.

8

CONCLUSION: ON CONTRADICTIONS

In studying a problem, we must shun subjectivity, one-sidedness and superficiality.

—MAO ZEDONG, "ON CONTRADICTION" (1937)

A NY ENDING to this book feels somewhat arbitrary, since the story of industrial development and pollution control in Futian continues to take many surprising twists and turns, as does China's path to economic development. During my last visit to the township, cadres were eager to bring me up to date on recent events. All three factories—the zinc smelter, the coking plant, and the coal-washing plant—remained closed in Futian. However, the zinc smelter, still owned and operated by Mr. Zhang and other members of his family, had reopened in Shilongba township, just a few kilometers west across the Yunnan provincial border. Shilongba offered a more hospitable climate for industrial investors: in addition to a three-year tax holiday for new enterprises, the township supplied factories with subsidized electrical power, which it purchased at low cost from Huaping county. Evidently, Shilongba also possessed another competitive advantage when it came to environmental enforcement; while uniform emissions standards were set by SEPA at the central-government level, enforcement was highly uneven in different political jurisdictions, and Huaping County EPB officials were willing to ignore the zinc smelter's violations.

After learning about the relocation of the zinc smelter, my initial plan was to continue studying the industrial development process across the provincial border, but this plan met with several obstacles. I held a formal meeting with the Shilongba Communist Party secretary and several other high-level township cadres to explain the nature of my research and gain their consent for the study. I found these officials relatively open to the idea of conducting research in their jurisdiction, particularly since Shilongba's comparatively well developed economy, which included a nonferrous metals smelting factory backed by South Korean investors, provided an interesting counterexample to Futian. While cadres in Futian struggled with the fiscal consequences of factory closures, Shilongba took in several hundred thousand yuan each month in factory revenues. When I asked the party secretary about the secret behind Shilongba's success, he coyly replied, "Our thinking has been liberated [*women de sixiang yijing jiefangle*]."

My luck changed, however, when I met with an obstinate fellow who led me to believe he was someone high in the chain of command at the foreign-owned nonferrous-metals factory. While I sat in the main office of the plant, observing workers donning hardhats and gasmasks as they punched in and out for their shifts, he made a series of calls on his cell phone to township cadres asking about my project. As he placed these phone calls, I watched as crucibles of molten ore were conveyed along the factory floor by a series of mechanical arms, noting how different this modern, high-tech operation was from Futian's small factories. The man was unable to reach the party secretary, who might have been able to help my cause. At length, through a strained grin, he told me, "I'm sorry, but Shilongba Township is not open yet [*hai meiyou kaifang*], so you just go back to Futian and do your research."

Stymied momentarily, and concerned somewhat about the fact that my official letter of support was from Sichuan University, which offered no protection for conducting research in Yunnan, I reluctantly returned to Futian. But I remained convinced that I had the support of the Communist Party secretary, whose decisions generally hold the most clout in rural China. I came back to Shilongba just a few days later, this time with a group of undergraduate students whom I had enlisted to help administer some of my household surveys. When we took a break for lunch at a small restaurant in the township center, however, the obstinate man returned, this time wearing a badge and pistol and identifying himself as the chief of police (*paichusuo suozhang*). Flanked by several uniformed officers, he told me to finish lunch and leave the township within the hour. Shaken by these events, and concerned for the safety of the students under my charge, I left Shilongba and never went back.

Back in Futian, the first thing I noticed was the lack of activity inside the industrial compound. Although the remnants of factories, coal, and metal tailings were scattered about, there were no workers, no families with their children, no plumes of smoke rising from the complex, and no noxious smell in the air. From the hilltop, however, I could still see the factories of

Shilongba and Daxing, located just kilometers away geographically but in a different world administratively, puffing away. The second thing I noticed was the boarded-up storefronts in Futian's open district in the center of the township. Some of the old businesses remained—the farmers' credit cooperative, the motorcycle mechanic, the fertilizer store, several small retail shops, and the restaurant owned by the parents of the little girl with malformed feet and developmental disabilities—but many others were permanently shuttered and locked.

CONTEXTUALIZING LOCAL ENVIRONMENTAL VALUES

One main focus of this book has been the environmental perceptions and values of people in Futian, as mediated through their lived experiences and told through their own narratives. I have suggested that ethnography is a tool well-suited to understanding the nuances of environmental perceptions and values. Large cross-national surveys, while allowing for interesting geographical and cultural comparisons, can gloss over much of the contextual richness in how people interact with their local environments through tilling the soil, working in the factory, and eating from the local food system. Because they are designed for comparative purposes, cross-cultural surveys also suffer from an inability to frame the issues in ways that are salient for people, a problem which ultimately compromises validity.

I had to chuckle when, picturing Futian in my mind, I reread one of the World Values Survey questions recently, which asked participants whether they would "pay 20 percent more for products, if it helped to protect the environment." Environmentally friendly products are of no concern for villagers, and, in any event, most people don't have 20 percent more money to spend on anything. If nothing else, I hope that anthropologists and other social scientists can help illuminate how environmental issues are embedded in specific cultural, economic, and political formations and how, consequently, they must be studied *within* those formations.

I found during the course of my study that many of Futian's residents were openly concerned about the air and water pollution caused by local factories. They were tired of experiencing chronic respiratory problems; fearful of the cumulative effects of contamination on the physical and cognitive health of their children; tired of scraping the soot from their belongings; and angry about damaged crops and contaminated irrigation water.

Their narratives about the effects of pollution on their daily lives stand in contrast to much of the accepted wisdom, grounded in ideas such as the postmaterialism thesis, that poor individuals and communities do not have the "luxury" of worrying about environmental problems (Dunlap and Mertig 1997; Inglehart 1995).

Villagers' concerns also tended to be linked with a widely shared dissatisfaction over liberal policy changes that effectively excluded them from the financial benefits of industry, especially wage-labor opportunities. Whereas township factories had been viewed as "communal capital" during the initial stages of reform, in the late reform period they became the property of outside investors, who intended to extract local resources and who had little interest in the long-term future of Futian. To a certain extent, industrial pollution constituted a moral concern for villagers since environmental degradation was linked with the radical upending of socialist ethics and a new emphasis on individual rather than collective benefit. Reform and Opening meant saying farewell to communal capital, the industrial infrastructure and assets that had been built over several decades by the sweat of villagers' brows.

Farmers, who because of their marginal economic position were most vulnerable to environmental degradation, bore the brunt of the environmental and economic damage caused by industrial pollution. In their narratives about the effects of pollution, farmers often cited the ethic of "eating from the mountain [*kao shan chi shan*]" as the basis for their environmental values, which had little in common with an abstract notion of "nature" or "wilderness" devoid of human influence. On the contrary, villagers' understandings of the environment, grounded in both cultural norms and economic dependence upon the land, had a fundamentally anthropocentric component to them. Their marginal position within the local political economy and their reliance upon the agrarian ecosystem were mirrored by a heightened awareness of environmental problems. Significantly, the vulnerability of farming households to environmental degradation was facilitated by liberal economic reforms like the Household Responsibility System, which re-created the family as a private economic unit. No longer shielded by the agricultural collective from the buffeting winds of economic and ecological change, farmers now occupy an extremely tenuous and risky place in the nation's economy. Their livelihoods are only as secure as the quality of their small plots of land. While I have told this story from the perspective of villagers in Futian, it is a story that likely resonates with most of China's 800 million peasants.

In one important sense, the anthropocentric nature of environmental values in Futian fits with what we know about environmental values in global context. In the United States and other developed countries of the West, for example, environmental movements began with pragmatic concerns over the occupational health and safety hazards posed by the dual processes of urbanization and industrialization during the early twentieth century (Gottlieb 1993). Only later did the movement proceed to more abstract concepts such as biodiversity conservation, holes in the ozone layer, or global climate change.

Of course, villagers' views about pollution in Futian were in no way homogeneous. Industrial laborers and investors exhibited what I have called "strategic risk repression," a capacity to ignore or downplay the hazards posed by pollution within the calculus of outsized remuneration from factory jobs. When industrial workers did recognize pollution as a problem, it was seen as a necessary byproduct of an eminently desirable goal: economic development.. This serves to illustrate the fact that environmental values are never politically neutral but are embedded within specific sociopolitical and economic structures. When we study environmental values, we cannot decontextualize them from other components of social, cultural, and economic life. The question "Who loses from industrial pollution?" cannot easily be separated from the question "Who gains from industrial profit?" For industrial workers and investors, and their families, the most salient concern was the threat of regulatory enforcement at the hands of Environmental Protection Bureau Officials, a fear that ultimately proved to be well founded.

ENVIRONMENTAL CIVIL SOCIETY WITH CHINESE CHARACTERISTICS

One of the key areas of inquiry for social scientists studying China today is the role of civil society—which I have defined as an intermediate realm between the family and the state characterized by collective action around shared values, interests, and goals—in public life (Chan 2005; Weller 2005). Examining the role of civil society in mediating and mitigating China's environmental crisis is an important component of this research agenda. In the midst of rapid social, political, and economic change, it is impossible to study environmental problems apart from issues such as public participation, trust, and regulatory transparency. Throughout this book, I have emphasized the fact that this presents us with a special problem in a place such as the People's Republic of China, where governmental respect for individual rights and views is poor and where citizens have little legal recourse to voice concerns about environmental problems.

Along these lines, recent scholarship has suggested that China today is experiencing dynamic environmentalism but no "environmental movement" (Stalley and Yang 2006:333). While there may be a certain truth to it, this conclusion does little more than raise a more difficult question: What's in a movement? If we adjust our scholarly lens to see only those movements that are formal and institutionalized, China indeed appears to lack a cogent environmental movement. But this overlooks the subtle, spontaneous, ad hoc collectivities that cohere around shared interests regarding environmental problems. In fact, these sorts of actions happen in rural China every day; official government statistics suggest that environmental protests number in the thousands each year (French 2005), and such public mobilization undoubtedly exerts pressure on environmental regulators and policymakers.

In Futian, villagers' concerns about their health, and their anger over perceived damages to their economic livelihoods, motivated them to get involved in cautious acts of environmental advocacy. They complained to the township government; they petitioned EPB officials; and they worked to alert the provincial media about factory emissions. I have suggested that we can use "pillars" and "processes" as heuristic devices to understand how environmental civil society in China both draws upon historical and cultural precedent and adapts to present circumstances through the use of various creative strategies. China's historical legacy of civic dissent—grounded in millennia-old concepts such as the mandate of heaven and solidified in mass movements such as the Cultural Revolution—is important in this regard. So, too, is the growing body of civil law that increasingly affords legal standing to individuals. In terms of processes, I have shown that environmental civil society in Futian centered on informal farmers' organizations, provincial media coverage of pollution, and cultural norms such as saving face.

Economic liberalization has also played a role in the growth of environmental civil society. The decoupling of industry from state, which has been a hallmark of Reform and Opening, has undoubtedly helped open the door for civil society mobilization in opposition to industrial pollution. In the early reform era, when the vast majority of rural factories were owned and managed collectively by township and village governments, criticism of the factory was ultimately criticism of the state, albeit the local state. This effectively suppressed civil action over industrial pollution since, on the one hand, many villagers benefited directly from factory revenue and jobs and, on the other hand, the political consequences of open opposition to the government would have made people think twice before defying government cadres. In the present era of privatization, however, the financial link between rural enterprises and local governments, though still important, has become less clear. This ambiguity makes it possible for more rigorous enforcement of environmental standards to take place, driven in part by concerned citizens with a growing knowledge of their rights.

The subtle burgeoning of environmental civil society in China today reminds us that no matter how solid a grip the central government appears to have on its citizens and on local officials, its control is never hegemonic; those at the margins possess creative strategies for affecting the political system. Nevertheless, a cautionary note is in order. Civil-society actors—whether in religious, social, or environmental arenas—occupy a tenuous space in Chinese politics. If the Futian case is indicative of general trends, greater openness in environmental decision making will come slowly and incrementally by working within, rather than overturning, existing power structures. Indeed, as the political scientists Kevin O'Brien and Lianjiang Li have rightly noted, "the roots of Chinese rightful resistance lie in the rich soil of central policy" (2006:5). The consolidated nature of political power in China may even prove over the long term to be an asset for environmental protection, once the central government and the populace decide to change course.[1]

SUSTAINABILITY AS CONTRADICTION?

Not long after the Renhe District EPB forced the closure of Futian's factories in 2003, Mr. Tian Weizhao, director of the Sichuan Province Environmental Protection Bureau, sat down for an interview with a staff writer from the SEPA-sponsored journal *Environment* to outline some of the challenges his agency faces. During the course of the interview, the director noted:

In my subjective opinion, the situation we find ourselves in, in terms of environmental protection, isn't that bad. But it's not ideal, either. Environmental protection in Sichuan exists in tension with a number of contradictions [*maodun*], [including] the contradiction between development and environmental protection.

(DONG 2005:56)

His remarks reference a common socialist trope about "contradictions [*ma-odun*]." In Marx's dialectical materialism, each economic mode of production—feudalist, capitalist, socialist, and communist—is seen to contain one or more untenable contradictions, which are the seeds of its ultimate demise. Following Marx and Lenin, Mao Zedong viewed these contradictions, and the struggle they engendered among the masses, as the engines that propelled humankind forward through history. In capitalist society, for example, the struggle to reconcile the interests of the bourgeoisie and the proletariat, in the midst of an economic system that enriched one by extracting value from the other, is the principal contradiction. Beginning with Marx, socialist thinkers believed such a contradiction would give rise to a revolution driven by class strife.

The insightful words of the Sichuan EPB director underscore contemporary China's most crucial, and perhaps most intractable, contradiction: the struggle for sustainability. The nation's liberal development path, felicitously termed "socialism with Chinese characteristics," has taken a heavy toll on the environment. China is now grappling with some of the worst environmental problems yet seen in human history, most of which stem from what the eco-Marxist scholar James O'Connor has called the "second contradiction of capitalism," whereby heavy natural resource extraction and unfettered capitalist industrial growth ultimately undermine the economic conditions necessary for production. The capitalist mode of production—which in my analysis includes the euphemistically named "socialism with Chinese characteristics"—destroys, rather than reproduces, the conditions necessary to sustain itself (O'Connor 1988).

For cadres and villagers in Futian, the struggle for sustainability opened up several important analytical pathways, which I have explored in this book. First, China's use of the international discourse of sustainability, which represents a significant shift away from the modernist-materialist environmental discourse of the Maoist era, is not merely empty rhetoric. SEPA, and its successor, the Ministry of Environmental Protection, bolstered by increased institutional capacity and a growing framework of law and policy, has grown into an agency with real regulatory teeth. Officials on the front lines of pollution control in the Renhe District EPB saw their enforcement duties as an extension of the growing national commitment to sustainability. These enforcement officials used increasingly strict tactics to accomplish their mandate, including the forced closure of industrial factories that were once the mainstay of the rural economy.

As a scholar, I have ambivalent feelings about what has happened in Futian. While my academic interest in environmental issues no doubt led me to approach my fieldwork with some of the biases common to the Western environmental movement, the complexity of Futian's story has caused me to question many of my own assumptions. The people who owned factories or spent long hours working in them—those who are part and parcel of China's new industrial might—are not simply faceless minions diabolically plotting about how best to destroy the environment. Like people everywhere, they are the products of specific sociopolitical formations. They are part of the throng of millions who followed the central government's exhortations to "leave the land, but not the countryside" and to "enter the factory, but not the city." In Futian, people like Mr. Zhang chose to open factories or to work in them because they could make a better living for their families; because they had lost secure access to farmland; and because they required a way to cope with incredible uncertainty in an economic system that could—and often did—change overnight.

In this sense, many of the people in this book—industrial entrepreneurs, township and village cadres, farmers—were confronted by structural forces beyond their control. In some ways, I came to view the factory owners and workers as victims of an ambiguous and unpredictable political economy in China, where the state gives every impression of "retreating" from many of its duties but retains the authority to selectively enforce laws and policies when it deems necessary. Futian's zinc smelter, coal-washing plant, and coking plant were only three among tens of thousands of small-scale rural enterprises that were shut down for environmental infractions during the late 1990s and early 2000s. In most cases, environmental officials closed these factories with little or no monetary compensation to the private investors who had been induced by central policy to purchase them.

If we are to fully understand the processes behind pollution enforcement and the consequences of these processes for economic and environmental sustainability, then we need to appreciate the complexity of governance in contemporary China and the divergent positions and interests of various state agencies in regards to the question of sustainable development. In Futian, the act of pollution enforcement revealed deep fissures between the township, district, and national governments, each of which, I have argued, advocated for its own model of sustainable development. The environmental enforcement process also highlighted the tension between rural industrial development and the agricultural sector of the economy upon which several hundred million Chinese peasants depend for their livelihoods. Questions regarding sustainable development ultimately involve, of course, a normative component: What should be sustained? The answer to that question will entail consequences for ecological integrity and human health, but also for economic growth, employment, and community development.

Most troubling perhaps is that the Futian case hits upon something that has not been properly resolved in any of the scientific literature on sustainable development: the undeniable contradiction between "sustaining" and "developing." The 1987 Brundtland Report, *Our Common Future*, and similar publications have provoked considerable discussion about the compatibility of economic growth and environmental protection. The discussion reverberates to this day. For example, the General Assembly of the United Nations, after holding a World Summit in New York in 2005, published an outcome document based on the contents of the summit. The document refers to economic development, social development, and environmental protection as the three "interdependent and mutually reinforcing pillars" of sustainable development (United Nations General Assembly 2005). But there is no meaningful attempt in the document to define or operationalize these concepts and no recognition that in some cases these "pillars" stand in fundamental opposition to one another. It is doubtful that such policy efforts, which stop short of acknowledging and addressing the unequal distribution of environmental threats inherent in the capitalist development paradigm, can result in genuine and productive change.

The sustainable-development movement shows a recognition on the part of scholars and policymakers of the intricate coupling of social and ecological systems, and to this extent it represents an analytical step forward, to be sure. But for Futian's cadres and villagers alike, achieving sustainability seemed akin to reconciling *wenbao* ("warmth and fullness") with *huanbao* ("environmental protection"). Having only recently emerged from the exceedingly difficult privations of the collective period, which some still darkly referred to as the "era of green shit," it was difficult to see the way forward. Rural enterprises, for all their grime, smoke, and stench, provided a tax base that could be used for carrying out vital development tasks such as building schools, hiring teachers, and bringing the "three connections" of roads, water, and electric power to the villagers. But the factories also undermined the agricultural livelihoods on which the majority of villagers depended.

The liberalization of China's economy and the concomitant privatization of the means of production benefited industrial investors while further marginalizing agrarian households. At the same time, the mounting environmental costs of industry—sullied rivers, hazy skies, and contaminated soils—were continually externalized and shrugged off onto those with the least capacity for defending themselves, especially the nation's farmers.

Economic equity and environmental sustainability are linked in intricate ways; ultimately, they cannot be achieved separately. In this sense, the power of the sustainability concept may lie in its capacity for bringing to the surface these contradictions and providing a discursive space within which they can be examined, discussed, and debated.

FUTIAN'S FUTURE, CHINA'S FUTURE

In the short term, cadres in Futian are not holding their breath while they wait for the golden age of industrial development to return. Many view a renewed reliance on small-scale market agriculture as the only available path to increase household incomes and restore the township's tax base. A recent economic-strategy document published by the township government focuses on the cultivation of commodity crops such as watermelon, honeydew, cantaloupe and mangos. Under the Household Responsibility System, villagers are free to choose what they grow on their fields, but the prospect of creating an agricultural economy of scale—in which the township's twenty-one small production cooperatives oversee cultivation and establish distribution networks in Panzhihua and beyond—is inducing many farmers to cooperate with their neighbors and kin to grow cash crops. The output of fruit crops alone has grown tenfold over the past few years. Futian has also received a subsidy from the central and provincial governments as part of the recently promulgated "new socialist countryside" program (*shehui zhuyi xin nongcun*); cadres are using the funds to construct a methane gas facility in Tangba Village, which converts household and animal wastes into a usable fuel source for household heating and cooking.

During my last visit to Futian, Li Fuming, the township's vice mayor, lamented over a cup of green tea that these improvements were enough to solve the "warmth and fullness problem [*wenbao wenti*]," but not nearly sufficient for villagers to attain "small comfort [*xiaokang*]." Nor could small-scale agriculture provide enough tax revenue to begin repayment on the township's 17 million yuan debt, which now represented the only tangible legacy of rural industrialization in Futian. From the window in his office in the open district, vice mayor Li gazed at the hilltop where the industrial compound had served as the hub of economic development for nearly two decades. "We have to use the resources at hand in order to promote development," he said slowly, "particularly now that industry is gone. We have water and good soil, and we're working to improve our agricultural output. We just have no good way to increase our tax revenue."

There are still many important aspects of China's current struggle with economic and environmental sustainability that require further exploration. As the PRC heads into a fourth decade of liberal economic reforms, bringing with it one-fifth of humanity, the magnitude of the environmental problems it faces—industrial pollution, deforestation, desertification, urban expansion, exploding consumption—suggests a renewed need for scholarly examination of the intricate linkages between ecological and social systems. In rural China, where rapid industrial development has produced pollution problems of immense proportion, a legacy of exposure to environmental toxins and damaged livelihoods remains. Left unchecked, these problems constitute an environmental, social, and public-health experiment whose long-term outcome is unknown but likely undesirable. In Futian, as elsewhere, cadres struggled to find a way to address the government's commitment to environmental sustainability without undoing the economic gains they had fought so hard to achieve.

Appendix

LIST OF CHINESE CHARACTERS

baijiu	白酒	*fen*	分
Bai zu	白族	*fengjian*	封建
biaozhun	标准	*fengshui*	风水
buzhu	补助	Futian	福田
chengbao tian	承包田	*Futian moshi*	福田模式
chi fan caizheng	吃饭彩正	*gaige kaifang*	改革开放
chi fan mei you	吃饭没有	*gang cun*	钢村
chi ku	吃苦	*gongye yuanqu*	工业园区
chunfen	春分	*guanli fei*	管理费
chuxinchang	出锌厂	Guantian Cun	官田村
cunweihui	村委会	*guanxi*	关系
Dai zu	傣族	*guizi*	贵子
danwei	单位	*hanhuale*	汉化了
datong	大同	*huanbao*	环保
Daxing Xiang	大兴乡	*huanbao wenti*	环保问题
dengji shiye lü	登记失业率	*huangshan*	黄山
di	地	*huanjing*	环境
diu mianzi	丢面子	*huanjing baohu*	环境保护
Dukou	渡口	*Huanjing baohu*	环境保护目标责
fa da yu quan	法大于权	*mubiao zeren zhi*	任制
fan tai zhi ge	钒钛之歌	*huanjing yingxiang*	环境影响评价法
fan zhengfu zuzhi	反政府组织	*pingjia fa*	
fazhan	发展	Huaping	华平
fazhan cai	发展才是硬道理	*hukou*	户口
shi ying daoli		*jihua shengyu*	计划生育
fazhan jinbu	发展进步报告	*jin*	斤
baogao		*jin chang, bu jin*	进厂不进城
fei zhengfu zuzhi	非政府组织	*cheng*	

jingshen	精神	*quanmian jianshe*	全民建设小康社会
Jingui Cun	金龟村	*xiaokang shehui*	
jin huang shiqi	金黄时期	*quanmin suoyou*	全民所有制
Jinjiang	金江	*zhi*	
Jinsha Jiang	金沙江	*ren*	人
jinxinchang	进锌厂	Renhe Qu	仁和区
jiti qiye	集体企业	*renmin daibiao*	人民代表大会
kaifa qu	开发区	*dahui*	
kao shan chi shan,	靠山吃山，靠水	*san fei*	三废
kao shui chi shui	吃水	*sange daibiao*	三个代表
kaohe zhibiao	考核指标	*shang you zhengce,*	上有政策，下有
kechixu fazhan	可持续发展	*xia you duice*	策
laobaixing	老百姓	*shan shui*	山水
laowai	老外	*shaoshu minzu*	少数民族
Liji	礼记	*shehui tuanti*	社会团体
litu, bulixiang	离土不离乡	*shehui wending*	社会稳定
liudong renkou	流动人口	*shehui zhuyi xin*	社会主义新农村
lübian shidai	绿便时代	*nongcun*	
maodun	矛盾	*shengchan dui*	生产队
meitan	煤炭	*shengtai huanjing*	生态环境
mianbao che	面包车	Shilongba	石龙坝
mianzi	面子	*shiwu xiao*	十五小
Miao zu	苗族	*shi yang jiu mu*	十羊九牧
minzu	民族	Shuitian	水田
mu	亩	*shuiyin*	水银
mubiao	目标	*si shi tong tang*	四世同堂
nongli	农历	Sichuan hua	四川话
nongye xue dazhai	农业学大寨	Tangba Cun	塘坝村
paichusuo	派出所	*Tao Te Ching*	道德经
Panzhihua	攀枝花	*tian*	田
Panzhihua Gangtie	攀枝花钢铁 (攀钢)	*tian*	天
(Pangang)		*tianming*	天命
pishuang	砒霜	*tian ren heyi*	天人合一
putonghua	普通话	*tongyi duo minzu*	统一多民族国家
qian	铅	*guojia*	
qingming	清明	*wai*	歪
qingnian dui	青年队	*waidiren*	外地人
qiufen	秋分	*wei renminbi fuwu*	为人民币服务
quan da yu fa	权大于法	*wei renmin fuwu*	为人民服务

wenbao wenti	温饱问题	*yimin cheng*	移民城
wenhua suzhi	文化素质	*yugong yi shan*	愚公移山
wenming shehui	文明社会	*yuan*	元
wuwei	无为	*zeren tian*	责任田
Wuzitian Cun	务子田村	*zhanlüe*	战略
xiagang	下岗	*zheng*	正
xiang wei qian	向为钱	*zhiqing*	知青
xiaokang	小康	*zhongguo tesede*	中国特色的社会
xiaomaibu	小卖部	*shehui zhuyi*	主义
xiao zu	小组	*zhua da fang xiao*	抓大放小
xibu da kaifa	西部大开发	*Zhuang zu*	壮族
xuyao kao ziji	需要靠自己	*ziran*	自然
Yanbian Xian	盐边 县	*zizhi qu*	自治区
yeman	野蛮	*zizhi xian*	自治县
Yi zu	彝族	*zizhi zhou*	自治州

NOTES

1. ENVIRONMENTAL VALUES, CIVIL SOCIETY, AND SUSTAINABILITY IN POST-REFORM CHINA

1. I use the *pinyin* system to Romanize all Mandarin Chinese words throughout the book, with the exception of some words or proper names that are better known in English by other spellings. A list of Chinese characters is included in the appendix.

2. Emerging research suggests that even residents of developing countries are concerned about "global" environmental problems, such as biodiversity conservation and climate change, and that global environmentalism is most likely a "collection of multiple movements beyond simply a North-South split" (see Brechin 1999:805–6).

3. In addition to the panzhihua tree, the region is also home to another celebrated floral species: *Cycas panzhihuaensis*, a 270-million-year-old cycad whose broad, serrated leaves resemble those of a fern.

4. During my most extensive period of fieldwork (2002 and 2003), the township comprised four villages: Wuzitian, Tangba, Guantian, and Jingui. In 2004, however, the administrative structure of Guantian village was dismantled when it was annexed by Wuzitian village; the township now consists of only three villages.

5. The *mu* is the traditional Chinese measure for land area. One *mu* is equal to 0.066 hectares or 0.165 acres.

2. THE DEVELOPMENT IMPERATIVE

1. The other two administrative areas are Miyi County to the northeast, and Yanbian County to the northwest. With 7,149 square kilometers of land under its jurisdiction, many government documents refer to Panzhihua as the "longest city in China."

2. The Chinese term for population, "*renkou*," is created by juxtaposing the character for "person" with the character for "mouth." China's population has numbered in the hundreds of millions since the Ming Dynasty, and bureaucrats have long recognized the strain on food resources created by too many "people's mouths."

3. The anthropologist Fei Xiaotong, who studied in Britain under Branislaw Malinowski, played a key role in the ethnic-identification project. The goal was to classify each *minzu* according to Marx's schema as primitive, slave, feudal, bourgeois-capitalist, socialist, or communist (Harrell 1995). The criteria for classifying *minzu* were borrowed from the Soviet Union's system, developed by Stalin, and included a common territory, language, mode of subsistence, and "psychological makeup." In practice, the process of affording recognized *minzu* status to some groups while denying it to others was highly politically charged since it was tied to the establishment of ethnic regional autonomy for governance in minority areas (Wang and Young 2006).

4. Tibet, for example is recognized as a provincial-level autonomous region (*zizhi qu*), but the potential for a secessionist movement causes the central government to keep a tight rein on the region. China has been reluctant to accord its minority nationalities "indigenous" status under such frameworks as the UN Convention on Biodiversity, which recognizes the rights of indigenous people for autonomy and self-determination.

5. The twelve administrative units included in the Great Western Opening policy are Chongqing municipality, Sichuan province, Guizhou province, Yunnan province, Tibet Autonomous Region, Shanxi province, Gansu province, Ningxia Hui Autonomous Region, Qinghai province, Xinjiang Uighur Autonomous Region, Inner Mongolia Autonomous Region, and Guangxi Zhuang Autonomous Region. One of the problems of focusing on resource extraction as a development strategy is that there is no guarantee that the financial proceeds will remain in the region. For example, the demand for electrical power in China's growing coastal cities is so great that much of the hydropower resources of western provinces gets sent eastward under a program called "Send Western Electricity East [*xi dian dong song*]," exacerbating regional economic disparities (Magee 2006).

6. On another occasion in the city of Chengdu, the syncretic nature of ethnic identity was illustrated to me in a very poignant way. My wife and I, along with several colleagues from the Sichuan Nationalities Research Institute, attended a New Year celebration hosted by members of the Yi community at an upscale hotel. Twenty or so young people in traditional dress performed a series of dances on a polished dance floor while, just steps away, some of their peers, dressed in holiday regalia, bowled a few frames in the hotel's bowling alley.

7. I am indebted to Stevan Harrell for this translation, which not only rhymes in English but also stays true to the tone and feeling of the original Chinese.

8. Excerpted from a speech given on June 12, 1987, entitled, "We Shall Speed Up Reform."

9. An argument for rural industrialization was made as early as the 1930s by Fei Xiaotong, who thought that rural industry held the key to increasing living standards in the countryside (Fei 1968:114–26). For an overview of rural industrial development during the Maoist period, see Perkins (1977).

10. Most demographers use the term "excess deaths" to include both deaths as a direct result of starvation and the compounded effects of the famine on fertility trends (Becker 1996). Judith Banister (1987) offers one of the most widely cited models of "excess deaths" during the Great Leap.

3. SAYING FAREWELL TO COMMUNAL CAPITAL

This chapter's first epigraph is taken from a reflective essay by a group of Sichuan economists following the sixteenth National Party Congress in 2002, at which Hu Jintao became the general secretary of the CCP (see Lin Ling et al. 2002).

1. Economic and cultural factors also undoubtedly play a role in China's declining rate of population growth during the reform era. Ethnographic evidence suggests that rural Chinese, eager to provide a higher standard of living for their families, increasingly choose to have only one child. It is extremely difficult to parse out the effects of family-planning policies from those of individual choice related to changing cultural preferences and economic necessity (Yan 2003).

2. The CPC has been so enamored with the results of Reform and Opening that it passed a resolution to amend the party's constitution at the Seventeenth National Congress, held in October 2007. The resolution is a formal acknowledgement by party officials that socialism with Chinese characteristics, the development path envisioned by Deng Xiaoping, has been a success. It reads, in part, "The congress affirms that since the introduction of the Reform and Opening Up Policy in the late 1970s, the Chinese Communists and the Chinese people have, in an indomitable enterprising spirit and with their spectacular practice of innovation, composed a new epic recounting the Chinese nation's ceaseless efforts to make progress and become stronger. . . . The fundamental reason behind all our achievements and progress since the Reform and Opening Up policy was introduced is, in the final analysis, that we have blazed a path of socialism with Chinese characteristics" (see Chinese Communist Party 2007).

3. One of the key questions in China's transition to a market-driven economy is how to resolve the issue of property rights. What happens when a factory, located on land held by the township or village government, is privatized? In Futian, the land-use fee (*tudi fei*) was used as a partial solution to this problem. While the fixed capital (infrastructure, equipment, etc.) was bought outright by private investors, the township government charged a fee for the use of its land.

4. The sociologist Lu Hanlong has analyzed how the *xiaokang* concept of the reform era was contextualized by values passed down from Chinese historical sources (see Lu 2000).

4. THE ENVIRONMENTAL COSTS OF PROGRESS

1. Estimates of economic losses from pollution vary widely. A recent World Bank study, which considered such factors as healthcare costs, lost work time, crop damage, infrastructural damage, and decreased productivity of forest and fishery resources, provides a comprehensive overview (see World Bank 2007).

2. Despite China's voracious appetite for coal, there are signs that leaders in Beijing are actively seeking to diversify the country's energy sources; in fact, China may reach its goal of relying on renewable energy sources for up to 15 percent of its energy needs by 2020 (Martinot and Li 2007). Given that hydropower is generally included in these analyses as "renewable," however, the environmental benefits of such an approach are questionable, since large hydroelectric dam facilities cause ecosystem fragmentation, displacement of human populations, species loss, changes in water quality and sediment load, and many other negative effects.

3. China's recent ranking as the world's leader in carbon dioxide emissions underscores the fact that its environmental problems present a global threat. I would also like to emphasize, however, that although my focus in this book is on local industrial production of goods for domestic use, a growing share of emissions stem from China's manufacturing of the world's consumer goods. Moreover, while China's total CO^2 emissions have surpassed the United States, its per capita emissions are still significantly below U.S. levels.

4. As I noted in the preface, the institutional structure of environmental protection has

changed considerably over the years. In of March 2008, SEPA was renamed the Ministry of Environmental Protection. Since most of my research was conducted during the SEPA years, and in an effort to avoid confusion, I refer to SEPA throughout this book.

5. There is continued debate about the relative effects on health from fine (< 2.5 micrometers, or $PM_{2.5}$) and ultrafine (< 1 micrometer, or PM_1) particles. Some epidemiological studies have found $PM_{2.5}$ to be more closely associated with health problems than PM_{10} (Schwartz, Dockery, and Neas 1996). I chose to monitor PM_{10} because data from the U.S. EPA calls into question the association between $PM_{2.5}$ and specific health outcomes and because China's SEPA has not set national standards for $PM_{2.5}$. The majority of epidemiological studies use PM_{10}, often referred to as "respirable particulate matter," as the exposure indicator.

6. There was one exception: sample period 4 (April 20–24, 2003) spanned only four days.

7. An independent-samples t-test was run. The difference between sample period 1 and sample period 4 was significant at $p < .01$.

8. For an interesting and accessible analysis of the many challenges of environmental health research, see Sylvia Noble Tesh's book *Uncertain Hazards* (Tesh 2000).

9. Although some of these projects have since gone forward, this event was widely viewed as the first serious test of the EIA law and a legitimation of SEPA's growing authority (see *Renminwang* 2005).

10. In this regard, villagers' experiences in Futian mirror those of citizens in other countries, including the United States, where public concern clashes with ambiguity, uncertainty, or denial on the part of environmental-health scientists and officials. The study of "contested illnesses" and their links to environmental contaminants is a growing research agenda in the social sciences (for example, see Brown 2007).

5. POLLUTION, PERCEPTIONS, AND ENVIRONMENTAL VALUES

1. For an examination of how European and North American streams of environmental thought have influenced policy in contemporary China and Taiwan, including their growing national-park and nature-preserve systems, see Robert Weller's *Discovering Nature: Globalization and Environmental Culture in China and Taiwan* (Weller 2006). Over the past several decades, China has seen the growth of hundreds of nature reserves that collectively cover more than 100 million hectares (Coggins 2003:14). In 2001 I had the privilege to visit Zhalong Nature Preserve in the northeastern province of Heilongjiang, where visitors can still witness the flight of the majestic red-crowned crane (*dandinghe*) amid vast expanses of marshland.

2. Under the process of Administrative Review in China, it is common for a local environmental-protection bureau to investigate damages caused by pollution and either mediate between parties or impose a decision, which may include ordering the polluter to pay compensation to the damaged party (see Palmer 1998:802).

3. Villagers in contemporary China, particularly those with a historical memory of the excesses and upheavals of the Maoist period, often view their local leaders with profound distrust. This extends to cadres' competency, loyalty to representing the villagers' views at higher administrative levels, and personal integrity and morality (Li 2004).

4. For an explanation of the use of scaled survey questions in anthropological research, see Ross (2004:105–7) and Bernard (1995:254–55).

5. For a complete explanation of the statistical methods used, including proportional odds modeling and Chi-square analysis, see Tilt (2006), O'Connell (2005), and Agresti (2002).

6. In the West, the prevailing ideas about "nature" and "wilderness" as fundamentally separate from human activity have been facilitated by the dual processes of industrialization and urbanization, which have alienated much of the populace from the environment. This alienation process is captured eloquently in an essay by the historian Richard White entitled, "'Are You an Environmentalist or Do You Work for a Living?'" (White 1995).

6. CIVIL SOCIETY AND THE POLITICS OF POLLUTION ENFORCEMENT

This chapter's epigraph is excerpted from a feature article in the German newspaper, *Der Spiegel*, entitled "The Chinese Miracle Will End Soon: Interview with China's Deputy Minister of the Environment" (*Der Spiegel* 2005).

1. Before 1998, the agency was known as the National Environmental Protection Administration. For an excellent look at the early evolution of the Chinese environmental bureaucracy, see Jahiel 1998.

2. I have avoided, perhaps conspicuously, using the term "environmental justice" in this book because many of the connotations of this term do not fit well with the Chinese case. First, as Robert Bullard reminds us in his recent edited volume, *The Quest for Environmental Justice*, one of the key axioms in environmental-justice struggles is: "race matters" (Bullard 2005:2). Much of the work on environmental justice in the United States, for example, has focused on how racial minorities and people of low socioeconomic status bear a disproportionate share of the health risks associated with environmentally polluting industries and activities. But although China exhibits great ethnic and cultural diversity, race is a trope that does not work so well in the Chinese context. Second, much of the recent work on environmental justice in the United States has been conducted under a growing legal framework supported by Executive Order 12898, which makes clear legal mandates on federal agencies to prevent unequal environmental burdens from falling on communities of color or poor communities. No similar statutes exist in China.

3. The case of Taiwan provides an interesting counterpoint to China. Both countries have experienced highly authoritarian governments for much of the twentieth century, and both have espoused a decidedly modernist view of the natural environment, promoting large-scale development and industrialization at great ecological cost. China and Taiwan are quite different, however, in terms of the social and political mechanisms citizens use for dealing with environmental ills such as pollution, nuclear power, and garbage disposal. In Taiwan, these mechanisms include citizen campaigns and protests that draw in Buddhist temples, kinship networks, township and village factions, and local thugs (Weller 2006).

4. Media coverage of the 2005 benzene spill was quite extensive. In cooperation with a Chinese student in public health, I analyzed a year's worth of transcripts on the incident from Xinhua News Agency reports and television broadcasts from *Jiaodian Fangtan*, one of the most popular news programs on China Central Television. The transcripts, which number more than one hundred, cover all aspects of the incident, including the effect on water quality, the implications for public health, and the political scandal surrounding the initial cover-up. While Xinhua News Agency, as

the official mouthpiece of the CCP, predictably avoided politically contentious issues and urged trust in government authorities, CCTV aired several news segments that exposed and condemned a major government cover-up regarding the environmental and human health effects of the benzene spill.

5. Furthermore, thousands of "Internet police" with a mandate from the Party to control and censor material on the Internet (felicitously dubbed the "Great Firewall of China") make it difficult to get opposing viewpoints in the Chinese media. For example, one rarely hears the Tibet issue talked about in terms other than those espoused by the central government: "the peaceful liberation of Tibet [*Xizang heping jiefang*]."

7. STRUGGLING FOR SUSTAINABILITY

1. One of the key questions in China's transition to a market economy is how to resolve the issue of property rights. What happens when a factory, located on communal land held by the township or village government, is privatized? In Futian, the land-use fee (*tudi fei*) is used as a partial solution to this problem. While the fixed capital (infrastructure, equipment, etc.) was purchased outright by private investors, the township government still received a royalty for the use of its land and maintained ultimate transfer rights on the land.

2. As Elizabeth Perry has noted, the use of the political regime's own words to support collective action at the grassroots level was "raised to an art form during the years of the Cultural Revolution" (Perry 2003:274). Such strategies remain important in reform-era China.

3. The three mountains likely refer to mountains considered sacred in the ancient Taoist tradition. One of the most significant of these, Mt. Huashan, has five peaks that are said to resemble the petals of a flower. Located in the north-central province of Shaanxi, it remains a popular destination for pilgrims. The poem is quoted from Judith Shapiro (2001:68).

4. It is worth noting that most of the important early global environmental initiatives, including the UN Conference on the Human Environment (1972), the World Commission on Environment and Development (1987), and the UN Conference on Environment and Development (1992), were grounded in neo-Malthusian assumptions. For the organizers of these events, who represented mostly Western industrialized nations, the primary concern was that unchecked population growth in the Third World would outstrip the earth's carrying capacity. In recent years, political organizing on the part of scientists and politicians in the developing world has resulted in a more balanced emphasis on population, consumption, energy, and technology in thinking about sustainable development. Agenda 21, for example, contains a section on encouraging environmentally responsible consumption.

5. Elizabeth Economy's book, *The River Runs Black*, provides a comprehensive look at China's environmental challenges. In regards to pollution control, she also finds that EPB interests and mandates run contrary to those of other government agencies and that when push comes to shove in politics, EPBs generally lose (Economy 2004:105–12).

6. In the scholarly literature on sustainability, there is a broad spectrum of opinions about the possibility of reconciling economic growth with sound environmental practices. For a useful overview of the sustainable-development concept, and some of its key polemics, see Robinson (2004).

7. The most recent survey undertaken by Chinese scientists in the lower reach of the Yangtze River in 2006 found no evidence of a viable freshwater dolphin population, leading to speculation that the species is in fact already extinct.

8. In addition to her environmental work, Dai has been actively involved in China's pro-democracy movement. Although many of her writings are harshly critical of the CCP, she has largely avoided sanction from the Party, probably because of family connections: Dai's adoptive father, Ye Jianying, played a major role in the socialist revolution and was considered one of the "ten great marshals" of the People's Liberation Army. He also held important political positions in the Chinese Communist Party through much of the latter half of the twentieth century.

8. CONCLUSION: ON CONTRADICTIONS

1. The current global financial crisis, which commenced as I finished the final revisions for this book, poses yet another challenge to the relationship between state and civil society in China. It remains to be seen how the CCP will handle the strife that will likely arise from slowed economic growth and rising unemployment.

WORKS CITED

Agrawal, Arun. 2005. *Environmentality: Technologies of Government and the Making of Subjects*. Durham, N.C.: Duke University Press.

Agresti, Alan. 2002. *Categorical Data Analysis*. New York: Wiley-Interscience.

Alford, William P., Robert P. Weller, L. Hall, K. R. Polenske, Y. Shen, and D. Zweig. 2002. "The Human Dimensions of Pollution Policy Implementation: Air Quality in Rural China." *Journal of Contemporary China* 11 (32): 495–513.

Anagnost, Ann. 1997. *National Past-Times: Narrative, Representation, and Power in Modern China*. Durham, N.C.: Duke University Press.

Anagnost, Ann. 2004. "The Corporeal Politics of Quality (Suzhi)." *Public Culture* 16 (2): 189–208.

Banister, Judith. 1987. *China's Changing Population*. Stanford, Calif.: Stanford University Press.

Barker, Holly M. 2003. *Bravo for the Marshallese: Regaining Control in a Post-Nuclear, Post-Colonial World*. Belmont, Calif.: Wadsworth.

Beck, Ulrich. 1992. *Risk Society: Towards a New Modernity*. London: Sage.

Becker, Jasper. 1996. *Hungry Ghosts: Mao's Secret Famine*. New York: Henry Holt.

Beckerman, Wilfred. 1998. "Sustainable Development: Is It a Useful Concept?" In *The Environmental Ethics and Policy Book*, ed. D. VanDeVeer and C. Pierce, 462–74. Belmont, Calif.: Wadsworth.

Bernard, H. Russell. 1995. *Research Methods in Anthropology: Qualitative and Quantitative Approaches*. 2nd ed. London: AltaMira Press.

Brechin, Steven R. 1999. "Objective Problems, Subjective Values, and Global Environmentalism: Evaluating the Postmaterialist Argument and Challenging a New Explanation." *Social Science Quarterly* 80 (4): 793–809.

Brook, J. R., T. F. Dann, and R. T. Burnett. 1997. "The Relationship Among TSP, PM10, PM2.5 and Inorganic Constituents of Atmospheric Particulate Matter at Multiple Canadian Locations." *Journal of the Air and Waste Management Association* 47:2–19.

Brown, Phil. 2007. *Toxic Exposures: Contested Illnesses and the Environmental Health Movement.* New York: Columbia University Press.

Bullard, Robert D. 2005. Introduction to *The Quest for Environmental Justice: Human Rights and the Politics of Pollution,* ed. R. D. Bullard, 1–16. San Francisco: Sierra Club Books.

Chan, Kin-Man. 2005. "The Development of NGOs Under a Post-Totalitarian Regime: The Case of China." In *Civil Life, Globalization, and Political Change in Asia: Organizing Between Family and State,* ed. R. P. Weller, 20–41. New York: Routledge.

Checker, Melissa. 2005. *Polluted Promises: Environmental Racism and the Search for Justice in a Southern Town.* New York: New York University Press.

Checker, Melissa. 2007. "'But I Know It's True': Environmental Risk Assessment, Justice, and Anthropology." *Human Organization* 66 (2): 112–24.

China Daily. 2000. "Headway in Curbing China's Worsening Environmental Pollution." Beijing. January 9.

China Environmental Yearbook. 2001. *Zhongguo Huanjing Nianjian* [China Environmental Yearbook]. Beijing: Zhongguo Huanjing Nianjian Chubanshe.

China Industrial Development Report. 2001. *Zhongguo Gongye Fazhan Baogao: Zhongguode Xin Shiji Zhanlue: Cong Gongye Daguo Zouxiang Gongye Qiangguo* [China Industrial Development Report: China's Challenge for the New Century: From and Industrially Strong Country to and Industrially Weak Country]. Beijing: Zhongguo Shehui Kexue Yuan Gongye Jingji Yanjiusuo [Industrial Economics Research Center, Chinese Academy of Social Sciences].

China TVE Yearbook Editorial Committee. 2004. *Zhongguo Xiangzhen Qiye Nianjian* [China Township and Village Enterprise Yearbook]. Beijing: Zhongguo Nongye Chubanshe.

China Township and Village Enterprises. 2002. "Ben Shijitou Ershinian Jianshe Xiaokang Shehui Si Da Mubiao" [The Four Major Goals in Establishing a Well-Off Society During the First Twenty Years of the Current Century]. *China Township and Village Enterprises* [Zhongguo Xiangzhen Qiye Zazhi] 192:4.

Chinese Communist Party. 2007. Resolution on Amendment to CPC Constitution. Beijing.

Chinese State Council. 1996. Guowuyuan Guanyu Huanjing Baohu Rogan Wentide Jueding [State Council Decisions Concerning Certain Environmental Protection Issues]. Beijing.

Chinese State Council. 2003. "Administrative Regulations on the Collection and Use of Pollutant Discharge Fees." *China Law and Policy.*

Chinese State Environmental Protection Administration. 2006. *Zhongguo Huanjing Zhuangkuang Gongbao* [Report on the State of the Environment in China]. Beijing: SEPA.

Chinese State Environmental Protection Administration. 1996. *Zhonghua Renmin Gongheguo Guojia Biaozhu: Huanjing Kongqi Zhiliang Biaozhun* [China's National Standards: Ambient Air Quality Standards]. Beijing: SEPA.

Chinese Statistical Bureau. 2001. *Zhongguo Laodong Tongji Nianjian* [China Labour Statistical Yearbook]. Beijing: Department of Population, Social Science, and Technology Statistics, National Bureau of Statistics.

Chinese Statistical Bureau. 2004. *China Statistical Yearbook.* Beijing: Chinese Statistical Bureau.

Chinese Statistical Bureau. 2005a. *Zhongguo Laodong Tongji Nianjian* [China Labor Statistical Yearbook]. Beijing: Department of Population, Social Science and Technology Statistics.

Chinese Statistical Bureau. 2005b. *Zhongguo Tongji Nianjian* [China Statistical Yearbook]. Beijing: Chinese Statistical Bureau.

Coggins, Christopher. 2003. *The Tiger and the Pangolin: Nature, Culture, and Conservation in China*. Honolulu: University of Hawaii Press.

Conable, Barber. 1987. *Address to the World Resources Institute*. Washington, D.C.: The World Bank.

Cronon, William. 1995. "The Trouble with Wilderness; or, Getting Back to the Wrong Nature." In *Uncommon Ground: Rethinking the Human Place in Nature*, ed. W. Cronon, 69–90. New York: Norton.

Dai, Qing, ed. 1994. *Yangtze! Yangtze!* Armonk, N.Y.: M. E. Sharpe.

Dai, Qing, ed. 1999. *The River Dragon Has Come*. Armonk, N.Y.: M. E. Sharpe.

Dasgupta, Susmita, Hua Wang, and David Wheeler. 1997. "Bending the Rules: Discretionary Pollution Control in China." The World Bank Development Research Group, Policy Research Working Paper 1761.

Day, Kristin. 2005. *China's Environment and the Challenge of Sustainable Development*. Armonk, N.Y.: M. E. Sharpe.

Deng, Xiaoping. 1994. *Selected Works of Deng Xiaoping*. Vol. 3. Beijing: Foreign Languages Press.

Der Spiegel. 2005. "The Chinese Miracle Will End Soon: Interview with China's Deputy Minister of the Environment." Berlin. March 7.

Derman, Bill, and Anne Ferguson. 2003. "Value of Water: Political Ecology and Water Reform in Southern Africa." *Human Organization* 62 (3): 277–88.

Dockery, Douglas W., Arden C. Pope III, Xiping Xu, John D. Spengler, James H. Ware, Martha E. Fay, Benjamin G. Ferris, and Frank E. Speizer. 1993. "An Association Between Air Pollution and Mortality in Six U.S. Cities." *New England Journal of Medicine* 329 (24): 1753–59.

Dong, Wenmao. 2005. "Huanjing Jueding Meilai: Sichuan Sheng Huanjing Baohu Ju Juzhang Tian Weizhao Fangtan Lu [The Future of Environmental Decision Making: Interview with Tian Weizhao, Director of Environmental Protection Bureau of Sichuan Province]." *Huanjing* [Environment] 310:56–57.

Douglas, Mary. 2002 [1966]. *Purity and Danger: An Analysis of the Concepts of Pollution and Taboo*. New York: Routledge.

Dunlap, Riley E., and Angela G. Mertig. 1997. "Global Environmental Concern: An Anomaly for Postmaterialism." *Social Science Quarterly* 78 (1): 24–29.

Dunlap, Riley E., George H. Gallup Jr., and Alec M. Gallup. 1993a. "Of Global Concern: Results of the Health of the Planet Survey." *Environment* 35 (9): 7–15, 33–39.

Dunlap, Riley E., George H. Gallup Jr., and Alec M. Gallup. 1993b. *Health of the Planet: Results of a 1992 International Environmental Opinion Survey of Citizens in Twenty-four Nations*. Princeton, N.J.: Gallup International Institute.

Economy, Elizabeth C. 2004. *The River Runs Black: The Environmental Challenge to China's Future*. Ithaca, N.Y.: Cornell University Press.

Edmonds, Richard Louis. 1998. "Studies on China's Environment." *The China Quarterly* 156:725–32.

Ester, P., H. Vinken, S. Simoes, and M. Aoyagi-Usui. 2002. *Culture and Sustainability: A Cross-National Study of Cultural Diversity and Environmental Priorities Among Mass Publics and Decision Makers*. Amsterdam: Dutch University Press.

Fei, Xiaotong. 1968. *China's Gentry: Essays in Rural-Urban Relations*. Chicago: University of Chicago Press.

Fei, Xiaotong. 1992 [1947]. *From the Soil: The Foundations of Chinese Society.* Trans. G. G. Hamilton and Z. Wang. Berkeley: University of California Press.

Ferris, Richard J., and Hongjun Zhang. 2005. "Environmental Law in the People's Republic of China: An Overview Describing Challenges and Providing Insights for Good Governance." In *China's Environment and the Challenge of Sustainable Development*, ed. Kristen A. Day, 66–101. Armonk, N.Y.: M. E. Sharpe.

Fortun, Kim. 2001. *Advocacy After Bhophal: Environmentalism, Disaster, New Global Orders.* Chicago: University of Chicago Press.

French, Howard W. 2005. "Anger in China Rises Over Threat to Environment." *New York Times.* July 19.

Futian Township People's Government. 2004. *Futian Zhen Minzu Zhuangkuang* [The Situation of Minority Nationalities in Futian Township]. Futian.

Gezon, Lisa L. 2006. *Global Visions, Local Landscapes: A Political Ecology of Conservation, Conflict, and Control in Northern Madagascar.* Lanham, Md.: AltaMira Press.

Gottlieb, Robert. 1993. *Forcing the Spring: The Transformation of the American Environmental Movement.* Washington, D.C.: Island Press.

Grantz, D. A., J. H. B. Garner, and D. W. Johnson. 2003. "Ecological Effects of Particulate Matter." *Environment International* 29:213–39.

Grossman, Gene M., and Alan B. Krueger. 1995. "Economic Growth and the Environment." *Quarterly Journal of Economics* 110:353–77.

Guan, Yanzhu. 2002. "Difang Huanjing Lifa Ying Zunning Kechixu Fazhan Yuanze [Local Environmental Legislation Should Abide by the Principles of Sustainable Development]." *Zhongguo Renkou, Ziyuan yu Huanjing* [China Population, Resources, and Environment] 12:18–16.

Guha, Ramachandra, and Juan Martinez-Alier, eds. 1997. *Varieties of Environmentalism: Essays North and South.* London: Earthscan.

Haenn, Nora. 2005. *Fields of Power, Forests of Discontent: Culture, Conservation, and the State in Mexico.* Tucson: University of Arizona Press.

Harrell, Stevan. 2001. *Ways of Being Ethnic in Southwest China.* Seattle: University of Washington Press.

Harrell, Stevan. 1995. "Introduction: Civilizing Projects and the Reaction to Them." In *Cultural Encounters on China's Ethnic Frontiers*, ed. S. Harrell, 3–36. Seattle: University of Washington Press.

Hefner, Robert. 1998. "On the History and Cross-Cultural Possibility of a Democratic Ideal." In *Democratic Civility: The History and Cross-Cultural Possibility of a Democratic Ideal*, ed. R. Hefner, 3–52. London: Transaction Press.

Ho, Peter. 2001. "Greening Without Conflict? Environmentalism, NGOs, and Civil Society in China." *Development and Change* 32 (5): 893–921.

Hoek, G., J. Schwartz, B. Groot, and P. Eilers. 1997. "Effects of Ambient Particulate Matter and Ozone on Daily Mortality in Rotterdam, The Netherlands." *Archives of Environmental Health* 52:455–63.

Inglehart, Ronald. 1995. "Public Support for Environmental Protection: Objective Problems and Subjective Values in Forty-three Societies." *Political Science and Politics* 28 (1): 57–72.

Inglehart, Ronald. 1997. *Modernization and Postmodernization: Cultural, Economic, and Political Change in Forty-three Societies.* Princeton, N.J.: Princeton University Press.

Ito, K., G. D. Thurston, and C. Hayes. 1993. "Associations of London, England, Daily Mortality with Particulate Matters, Sulphur Dioxide, and Acidic Aerosol Pollution." *Archives of Environmental Health* 48:213–20.

Jahiel, Abigail R. 1998. "The Organization of Environmental Protection in China." *The China Quarterly* 156:757–87.

Jing, Jun. 2000. "Environmental Protests in Rural China." In *Chinese Society: Change, Conflict and Resistance*, ed. E. J. Perry and M. Selden, 143–60. London: Routledge.

Jinside Panzhihua Bianweihui [Golden Panzhihua Editorial Committee], ed. 1990. *Jinside Panzhihua* [Golden Panzhihua]. Chengdu: Sichuan Sheng Kexue Jishu Chubanshe [Sichuan Province Science and Technology Press].

Kan, Haidong, and Bincheng Chen. 2003. "Air Pollution and Daily Mortality in Shanghai: A Time-Series Study." *Archives of Environmental Health* 58 (6): 360–67.

Khan, Azizur Rahman, and Carl Riskin. 1998. "Income and Inequality in China: Composition, Distribution, and Growth of Household Income, 1988 to 1995." *The China Quarterly* 154:221–53.

Kottak, Conrad. 1999. "The New Ecological Anthropology." *American Anthropologist* 101 (1): 23–35.

LaFargue, Michael. 2001. "'Nature' as Part of Human Culture in Daoism." In *Daoism and Ecology: Ways Within a Cosmic Landscape*, ed. N. J. Girardot, J. Miller, and X. Liu, 45–60. Cambridge, Mass.: Harvard University Press.

Lai, Honyi H. 2002. "China's Western Development Program: Its Rationale, Implementation, and Prospects." *Modern China* 28 (4):432–66.

Lee, Yok-Shiu F. 2005. "Public Environmental Consciousness in China: Early Empirical Evidence." In *China's Environment and the Challenge of Sustainable Development*, ed Kristen A. Day, 35–65. Armonk, N.Y.: M. E. Sharpe.

Leopold, Aldo. 1949. *A Sand County Almanac, and Sketches Here and There.* New York: Oxford University Press.

Li, Hongbin, and Scott Rozelle. 2003. "Privatizing Rural China: Insider Privatization, Innovative Contracts and the Performance of Township Enterprises." *The China Quarterly* 176:981–1005.

Li, Lianjiang. 2004. "Political Trust in Rural China." *Modern China* 30 (2): 228–58.

Li, Yongxiang, and Bryan Tilt. 2007. "In Search of Solvency: Changing Agricultural Governance in an Ethnic Minority Autonomous Region in Southwest China." *International Journal of Agricultural Resources, Governance, and Ecology* 6 (6):626–41.

Li Xingxing. 1995. *Panxi Lieshangde Shengchangdian: Panzhihua Jingji Fushen yu Minzu Guanxi* [Development in the Panzhihua-Xichang Valley: Panzhihua Economic Development and Ethnic Relations]. Chengdu: Sichuan Minzu Chubanshe [Sichuan Ethnic Minority Press].

Lin, Ling, Kendang Du, Shuiping Hou, Shouqi Du, Houming She, and Gang Yang. 2002. "Yu Shidai Jujin Shixin Quanmian Jianshe Xiaokang Shehui: Sichuan Sheng Zheming Jingji Xuejia Tan Xuexi Dangde Shiliu Da Baogaode Tihui [Keep Pace with the Times, Blaze and Create Something New, Build a Well-Off Society in an All-Around Way: What Some Sichuan Economists Learned from the Report on the Sixteenth National Party Congress of the CCP]." *Jingji Zhidu Gaige* [Reform of Economic System] 117:6–7.

Lin Qingsong and William A. Byrd. 1994. *Zhongguo Xiangzhen Qiyede Lishi Xingjuqi: Jieshao, Fazhan yu Gaige* [A Historical Examination of China's Township and Village Enterprises: Beginnings, Development, and Reform]. Oxford: Oxford University Press.

Liu, Jianguo, and Jared Diamond. 2005. "China's Environment in a Globalizing World." *Nature* 435:1179–86.

Liu, Juliana. 2007. "Pollution Turns China Village Into Cancer Cluster." Reuters. Beijing. September 20.

Liu, S., M. R. Carter, and Y. Yao. 1998. "Dimensions and Diversity of Property Rights in

Rural China: Dilemmas on the Road to Further Reform." *World Development* 26 (10): 1798–1806.

Liu, Yanhua, and Yanchun Zhou. 2001. *Zhongguo Ziyuan Huanjing Xingshi yu Kechixu Fazhan* [The State of China's Natural Resources and Sustainable Development]. Beijing: Jingji Kexue Chubanshe [Economic Sciences Press].

Lo, Carlos Wing Hung, and Sai Wing Leung. 2000. "Environmental Agency and Public Opinion in Guangzhou: The Limits of a Popular Approach to Environmental Governance." *The China Quarterly* 163:677–704.

Lowe, Celia. 2006. *Wild Profusion: Conservation in an Indonesian Archipelago*. Princeton, N.J.: Princeton University Press.

Lu, Hanlong. 2000. "To Be Relatively Comfortable in an Egalitarian Society." In *The Consumer Revolution in Urban China*, ed. D. Davis, 124–41. Berkeley: University of California Press.

Ma, Xiaoying, and Leonard Ortolano. 2000. *Environmental Regulation in China: Institutions, Enforcement, and Compliance*. Lanham, Md.: Rowman and Littlefield.

Magee, Darrin. 2006. "Powershed Politics: Hydropower and Interprovincial Relations Under Great Western Development." *The China Quarterly* 185:23–41.

Mao, Zedong. 1966. *Nongken* [Agricultural Reclamation]. Volume 6.

Mao, Zedong. 1971. *Selected Readings from the Works of Mao Tsetung*. Beijing: Foreign Languages Press.

Martinez-Alier, Juan. 1995. "Commentary: The Environment as a Luxury Good, or 'Too Poor to be Green.'" *Ecological Economics* 13 (1):1–10.

Martinez-Alier, Juan. 2002. *The Environmentalism of the Poor: A Study of Ecological Conflicts and Valuation*. Cheltenham, U.K.: Edward Elgar.

Martinot, Eric, and Junfeng Li. 2007. *Powering China's Development: The Role of Renewable Energy*. Washington, D.C.: Worldwatch Institute.

Meisner, Maurice. 1996. *The Deng Xiaoping Era: An Inquiry Into the Fate of Chinese Socialism, 1978–1994*. New York: Hill and Wang.

Meng, Lang. 1999. *Huanjing Baohu Shidian* [Case Studies in Environmental Protection]. Changsha: Hunan Daxue Chubanshe [Hunan University Press].

Mol, Arthur P. J., and Neil T. Carter. 2006. "China's Environmental Governance in Transition." *Environmental Politics* 15 (2): 149–70.

Muldavin, Joshua S. S. 1996. "The Political Ecology of Agrarian Reform in China: The Case of Heilongjiang Province." In *Liberation Ecologies: Environment, Development, Social Movements*, ed. R. Peet and M. Watts, 227–59. London: Routledge.

Naughton, Barry. 1988. "The Third Front: Defense Industrialization in the Chinese Interior." *The China Quarterly* 115:351–86.

Naughton, Barry. 1992. "Implications of the State Monopoly Over Industry and Its Relaxation." *Modern China* 18 (1): 14–41.

New York Times. 2005. "China Fires Water Monitor." *New York Times*. December 3.

Ngai, Pun. 2005. *Made in China: Women Factory Workers in a Global Workplace*. Durham, N.C.: Duke University Press.

O'Brien, Kevin J., and Lianjiang Li. 2004. "Suing the Local State: Administrative Litigation in Rural China." *The China Journal* 51:75–96.

O'Brien, Kevin J., and Lianjiang Li. 2006. *Rightful Resistance in Rural China*. Cambridge: Cambridge University Press.

O'Connell, Ann A. 2005. *Logistic Regression Models for Ordinal Response Variables*. Thousand Oaks, Calif.: Sage.

O'Connor, James. 1988. "Capitalism, Nature, Socialism: A Theoretical Introduction." *Capitalism, Nature, Socialism* 1 (1): 11–38.

Oi, Jean C. 2001. "Realms of Freedom in Post-Mao China." In *Realms of Freedom in Modern China*, ed. W. C. Kirby, 264–84. Stanford, Calif.: Stanford University Press.

Oi, Jean C. 2005. "Patterns of Corporate Restructuring in China: Political Constraints on Privatization." *The China Journal* 53:115–36.

Osorio, Leonardo Alberto Rios, Manuel Ortiz Lobato, and Xavier Alvarez Del Castillo. 2005. "Debates on Sustainable Development: Towards a Holistic View of Reality." *Environment, Development, and Sustainability* 7:501–18.

Palmer, Michael. 1998. "Environmental Regulation in the People's Republic of China: The Face of Domestic Law." *The China Quarterly* 156:788–808.

Panzhihua City Statistical Bureau. 2004. *Panzhihua Xiangzhen Nianjian* [Panzhihua Township Yearbook]. Panzhihua: Panzhihua Shi Xiangzhen Pingjia Kaohe [Panzhihua City Office of Township Appraisal and Inspection, 1999–2004].

Panzhihua Municipal Government. 2003. *Zhongguo Panzhihua* [Panzhihua, China]. Panzhihua: Panzhihua Shi Renmin Zhengfu Xinwen Bangongshi [Information Office of the Municipal Government of Panzhihua].

Parris, Thomas M., and Robert W. Kates. 2003. "Characterizing and Measuring Sustainable Development." *Annual Review of Environment and Resources* 28:559–86.

Pei, Minxin. 2003. "Rights and Resistance: The Changing Contexts of the Dissident Movement." In *Chinese Society: Change, Conflict, and Resistance*, ed. E. J. Perry and M. Selden, 23–46. New York: Routledge.

Pellow, David Naguib, and Robert J. Brulle. 2005. "Power, Justice, and the Environment: Toward Critical Environmental Justice Studies." In *Power, Justice, and the Environment: A Critical Appraisal of the Environmental Justice Movement*, ed. D. N. Pellow and R. J. Brulle, 1–19. Cambridge, Mass.: MIT Press.

Peluso, Nancy Lee. 1994. *Rich Forests, Poor People: Resource Control and Resistance in Java.* Berkeley: University of California Press.

Peng, Chaoyang, et al. 2002. "Urban Air Quality and Health in China." *Urban Studies* 39 (12): 2283–99.

Perkins, Dwight. 1977. *Rural Small-Scale Industry in the People's Republic of China.* Berkeley: University of California Press.

Perry, Elizabeth J. 2003. " 'To Rebel Is Justified': Cultural Revolution Influences on Contemporary Chinese Protest." In *The Chinese Cultural Revolution Reconsidered: Beyond Purge and Holocaust*, ed. K.-Y. Law, 262–81. New York: Palgrave MacMillan.

Petryna, Adriana. 2002. *Life Exposed: Biological Citizens After Chernobyl.* Princeton, N.J.: Princeton University Press.

Primbs, T., S. L. Simonich, D. Schmedding, G. Wilson, D. Jaffe, A. Takami, S. Kato, S. Hatakeyama, and Y. Kajii. 2007. "Atmospheric Outflow of Anthropogenic Semi-Volatile Organic Compounds from East Asia in Spring 2004." *Environmental Science and Technology* 41 (10): 3551–58.

Qian, Zhengmin, Robert S. Chapman, Qiuxue Tian, and Yan Chen. 2000. "Effects of Air Pollution on Children's Respiratory Health in Three Chinese Cities." *Archives of Environmental Health* 55 (2): 126–33.

Qu, Geping. 2002. "Guanzhu Shengtai Anquan Zhiyi, Shengtai Huanjing Wenti yu Jingcheng Wei Guojia Anquande Remen Huati [The Problems of Ecological Environment Have Become a Popular Subject of National Safety]." *Huanjing Baohu* [Environmental Protection] 295:3–4.

Ren, Hongwei, and Gong Li. 2002. "Xiangzhen Qiye Huanjing Wurande Jingji Fenxi [An Economic Analysis of the Environmental Pollution of Township and Village Enterprises]." *Zhongguo Xiangzhen Qiye* [China Township and Village Enterprises] 1 (181): 34.

Renminwang. 2005. "30 Ge Weifa Kaigong Xiangmu Mingdan Gongbu 'Huanping Fengbao' Guajin Huanbao Xian [Names of Thirty Illegally Started Projects Announced, 'Tempest of Environmental Review' Tightens the String of Environmental Protection]." *Renminwang*. Beijing. January 18.

Riskin, Carl, Renwei Zhao, and Shi Li. 2001. *China's Retreat from Equality: Income Distribution and Economic Transition*. Armonk, N.Y.: M. E. Sharpe.

Robinson, John. 2004. "Squaring the Circle? Some Thoughts on the Idea of Sustainable Development." *Ecological Economics* 48:369–84.

Rosen, Daniel H., and Trevor Houser. 2007. "China Energy: A Guide for the Perplexed." Policy Brief, Peterson Institute for International Economics.

Ross, Norbert. 2004. *Culture and Cognition: Implications for Theory and Method*. Thousand Oaks, Calif.: Sage.

Rossabi, Morris. 2004. Introduction to *Governing China's Multiethnic Frontiers*, ed. M. Rossabi, 3–18. Seattle: University of Washington Press.

Rozelle, Scott, Loren Brant, Guo Li, and Jikun Huang. 2005. "Land Tenure in China: Facts, Fictions, and Issues." In *Developmental Dilemmas: Land Reform and Institutional Change in China*, ed. P. Ho, 121–50. New York: Routledge.

Samet, Jonathan M., Francesca Dominici, Frank C. Curriero, Ivan Coursac, and Scott L. Zegger. 2000. "Fine Particulate Air Pollution and Mortality in Twenty U.S. Cities." *The New England Journal of Medicine* 343 (24): 1742–48.

Schwartz, J., D. Dockery, and L. Neas. 1996. "Is Daily Mortality Associated Specifically with Fine Particulates?" *Journal of Air and Waste Management Association* 46:2–14.

Scott, James C. 1985. *Weapons of the Weak: Everyday Forms of Peasant Resistance*. New Haven, Conn.: Yale University Press.

Selden, Mark. 1998. "House, Cooperative, and State in the Remaking of China's Countryside." In *Cooperative and Collective in China's Rural Development: Between State and Private Interests*, ed. E. B. Vermeer, F. N. Pieke, and W. L. Chong, 17–45. Armonk, N.Y.: M. E. Sharpe.

Seligman, Adam B. 1992. *The Idea of Civil Society*. New York: Free Press.

Shapiro, Judith. 2001. *Mao's War Against Nature: Politics and the Environment in Revolutionary China*. Cambridge: Cambridge University Press.

Short, Philip. 1999. *Mao: A Life*. New York: Henry Holt.

Sichuan Province Panzhihua City Editorial Committee. 1995. *Panzhihua Shizhi* [Panzhihua City Record]. Chengdu: Sichuan Kexue Jishu Chubanshe [Sichuan Science and Technology Press].

Sichuan Statistical Bureau. 2002. *Sichuan Sheng Tongji Nianjian* [Sichuan Province Statistical Yearbook]. Chengdu: Sichuan Statistical Bureau.

Sichuan Statistical Bureau. 2005. *Sichuan Statistical Yearbook* [Sichuan Sheng Tongji Nianjian]. Chengdu: Sichuan Statistical Bureau.

Smil, Vaclav. 2004. *China's Past, China's Future: Energy, Food, Environment*. New York: Routledge.

Solinger, Dorothy J. 1995. "China's Urban Transients in the Transition from Socialism and the Collapse of the Communist 'Urban Public Goods Regime.'" *Comparative Politics* 27 (2): 127–46.

Solinger, Dorothy J. 1997. "The Impact of the Floating Population on the Danwei: Shifts in the Patterns of Labor Mobility Control and Entitlement Provision." In *Danwei: The Changing Chinese Workplace in Historical and Comparative Perspective*, ed. Xiaobo Lu and Elizabeth J. Perry, 195–222. Armonk, N.Y.: M. E. Sharpe.

Stalley, Phillip, and Dongning Yang. 2006. "An Emerging Environmental Movement in China?" *The China Quarterly* 186:333–56.

Starr, John Bryan. 1997. *Understanding China: A Guide to China's Political Economy, History, and Political Structure*. New York: Hill and Wang.

Sun, Changmin. 2000. "The Floating Population and Internal Migration in China." In *The Changing Population of China*, ed. X. Peng and Z. Guo, 179–91. Oxford: Blackwell.

Tesh, Sylvia Noble. 2000. *Uncertain Hazards: Environmental Activists and Scientific Proof.* Ithaca, N.Y.: Cornell University Press.

Tilt, Bryan. 2006. "Perceptions of Risk from Industrial Pollution in China: A Comparison of Occupational Groups." *Human Organization* 65 (2): 115–27.

Tilt, Bryan. 2007. "The Political Ecology of Pollution Enforcement in China: A Case from Sichuan's Rural Industrial Sector." *The China Quarterly* 192:915–32.

Tilt, Bryan. 2008. "Smallholders and the 'Household Responsibility System': Adapting to Institutional Change in Chinese Agriculture." *Human Ecology* 36 (2): 189–99.

Tilt, Bryan, and Pichu Xiao. 2007. "Industry, Pollution, and Environmental Enforcement in Rural China: Implications for Sustainable Development." *Urban Anthropology and Studies of Cultural Systems and World Economic Development* 36 (1–2): 115–43.

Tsing, Anna Lowenhaupt. 2005. *Friction: An Ethnography of Global Connection*. Princeton, N.J.: Princeton University Press.

Tu, Weiming. 1998. "The Continuity of Being: Chinese Visions of Nature." In *Confucianism and Ecology: The Interrelation of Heaven, Earth, and Humans*, ed. M. E. Tucker and J. Berthrong, 105–22. Cambridge, Mass.: Harvard University Press.

United Nations Development Program. 2007. *Human Development Report 2007/2008: Fighting Climate Change: Human Solidarity in a Divided World*. New York: United Nations.

United Nations General Assembly. 2005. *World Summit Outcome*. New York: United Nations.

Vermeer, Edward B. 1998. "Industrial Pollution in China and Remedial Policies." *The China Quarterly* 156:952–85.

Vidal, John and David Adam. 2007. "China Overtakes U.S. as World's Biggest CO_2 Emitter." *The Guardian*. London. June 19.

Wan, Siji. 1999. *Daqi Wuran Zhili Gongcheng* [Atmospheric Pollution Control Engineering]. Beijing: Gaodeng Jiaoyu Chubanshe [Higher Education Press].

Wang, Hua, Nlandu Mamingi, Benoit Laplante, and Susmita Dasgupta. 2003. "Incomplete Enforcement of Pollution Regulation: Bargaining Power of Chinese Factories." *Environmental and Resource Economics* 24:245–62.

Wang, Hua, and David Wheeler. 2005. "Financial Incentives and Endogenous Enforcement in China's Pollution Levy System." *Journal of Environmental Economics and Management* 49 (1): 174–96.

Wang, Jianmin, and John A. Young. "Applied Anthropology in China." *NAPA Bulletin* 25:70–81.

Wang, Shaoguang, and Angang Hu. 1999. *The Political Economy of Uneven Development: The Case of China*. Armonk, N.Y.: M. E. Sharpe.

Wang, Shinliang, Kongxiu Wang, and Sharong Wang. 2000. *Kechixu Fazhan Gailun* [An

Introduction to Sustainable Development]. Jinan: Shandong Renmin Chubanshe [Shandong People's Press].

Weller, Robert P. 2005. "Introduction: Civil Institutions and the State." In *Civil Life, Globalization, and Political Change in Asia: Organizing Between Family and State*, ed. R. P. Weller, 1–19. New York: Routledge.

Weller, Robert P. 2006. *Discovering Nature: Globalization and Environmental Culture in China and Taiwan*. Cambridge: Cambridge University Press.

West, Paige. 2006. *Conservation Is Our Government Now: The Politics of Ecology in Papua New Guinea*. Durham, N.C.: Duke University Press.

West, Paige, and Dan Brockington. 2006. "An Anthropological Perspective on Some Unexpected Consequences of Protected Areas." *Conservation Biology* 20 (3): 609–16.

Wheeler, David, Hua Wang, and Susmita Dasgupta. 2003. "Can China Grow and Safeguard Its Environment: The Case of Industrial Pollution." In *How Far Across the River? Chinese Policy Reform at the Millennium*, ed. N. C. Hope, D. T. Yang, and M. Y. Li, 353–88. Stanford, Calif.: Stanford University Press.

White, Lynn Jr. 1967. "The Historical Roots of Our Ecological Crisis." *Science* 155:1203–7.

White, Richard. 1995. "'Are You an Environmentalist or Do You Work for a Living?': Work and Nature." In *Uncommon Ground: Toward Reinventing Nature*, ed. W. Cronon, 171–86. New York: Norton.

Whiting, Susan. 2000. *Power and Wealth in Rural China: The Political Economy of Institutional Change*. New York: Cambridge University Press.

World Bank. 1997. *China's Environment in the New Century: Clear Water, Blue Skies*. Washington, D.C.: World Bank.

World Bank. 2007. *The Cost of Pollution in China: Economic Estimates of Physical Damages*. Washington, D.C.: World Bank.

World Commission on Environment and Development. 1987. *Our Common Future*. Oxford: Oxford University Press.

Xi, Xiaolin, and Xu Qinghua, eds. 1999. *Zhongguo Gongzhong Huanjing Yishi Diaocha* [A Survey on China's Public Environmental Consciousness]. Beijing: Zhongguo Huanjing Kexue Chubanshe [China Environmental Science Press].

Xu, Shaoyi, Dogian Yu, Libin Jing, and Xiping Xu. 2000. "Air Pollution and Daily Mortality in Shenyang, China." *Archives of Environmental Health* 55 (2): 115–24.

Xue, Lan, Udo E. Simonis, and Daniel J. Dudek. 2007. "Environmental Governance for China: Major Recommendations of a Task Force." *Environmental Politics* 16 (4): 669–76.

Yan, Yunxiang. 2003. *Private Life Under Socialism: Love, Intimacy, and Family Change in a Chinese Village, 1949–1999*. Stanford, Calif.: Stanford University Press.

Yan, Yunxiang. 1996. *The Flow of Gifts: Reciprocity and Social Networks in a Chinese Village*. Stanford, Calif.: Stanford University Press.

Yang, Guobin. 2005. "Environmental NGOs and Institutional Dynamics in China." *The China Quarterly* 181:46–66.

Yang, Mayfair. 1994. *Gifts, Favors, and Banquets: The Art of Social Relationships in China*. Ithaca, N.Y.: Cornell University Press.

Zhang, Li. 2001. *Strangers in the City: Reconfigurations of Space, Power, and Social Networks Within China's Floating Population*. Stanford, Calif.: Stanford University Press.

Zhang, Ze. 2005. "Shiyi Wu Guihua Xiade Huanjing Touzi [Environmental Investment in the Eleventh Five-Year Plan]." *Huanjing* [Environment] 310:38–40.

Zhou, Jigang. 2006. "The Rich Consume and the Poor Suffer the Pollution." Zhong Wai Duihua [China Dialogue]. Beijing. October 27.

INDEX

Global Environmental Survey, 6
government mistrust, 94–95
gravimetric analysis, 73
great equality, age of, 62
Great Leap Forward: industrialization in, 37, 86; results of, 38
Great Western Opening policy, 32, 146–47; administrative units included in, 168n5; goal of, 33, 147; Little Hu on, 146–47
green research agenda, 16–17
guanxi, 40, 47, 56, 59

"Hanification," 33
"harmony between people and nature," 5
Harrell, Stevan, xiv
Health of the Planet survey, 1992, 6
Ho, Peter, 124
household-registration system. *See hukou*
Household Responsibility System, 2, 3, 26–27, 28–29, 91; farmers view on, 29–30; launch of, 38; recreating family as private economic unit, 154; as upright, 95
Huai River, 106
huanbao, wenbao vs., 142–43, 160
Hu Jintao, 63
hukou, 54–55; blue-stamp, 55; given to migrant workers, 132; legalization of, 54; relaxation of, 55; urban public goods regime and, 54
hydroelectric power station, 92; costs related to, 93; potential returns from, 92, 135; Zhang Huachao on, 93

illnesses, contested, 170n10
Indonesia, 17, 125
industrialization, 36–40; Fei Xiaotong on, 168n9; in Great Leap Forward, 37, 86; monetary benefits of, 90–91; problems threatening, 68; without urbanization, 54–55
Inglehart, Ronald, 5–6
insider privatization, 59
Internet, 10, 126; censoring, 172n5
iron rice bowl. *See* employment, guaranteed lifetime

Japan, 68
Jiang Zemin, 146; on Great Western Opening policy, 32; on strategy of sustainability, 138; on *xiaokang*, 11, 63, 138–39
Jiaodian Fangtan, 171n4
Jingui Village, 93, 94
Jinsha River, 13

kechixu fazhan. See development, sustainable
Korea, 68

land: collective ownership of, 91; contract, 27; dry agricultural, 27; intercropping, 28; irrigated paddy, 27; Mao Zedong and reform of, 26; parcelization, 27–28; relay cropping, 28; responsibility, 27; tenure system, 91; types of, 27
land, waste, 27, 28; migrant workers farming on, 132
language barrier, xiii
Laozi, 87
law enforcement, 5, 73; of air quality standards, 114; ambiguity in, 112; civil-society factors and, 119–20; exposé on, 114–15; fiscal concerns over, 118; prioritizing, 118; tools, 120
lead, 80
Leopold, Aldo, 86
Liang Congjie, 123
Lianjiang Li, 9, 122–23, 157
Li Fuming, 161
Li Hong, 61; on factory closures, 130
Li Jiejie, xi, 133
Lin Ling, 44
Little Hu, 57–58, 64; on factory closures, 130; on Great Western Opening, 146–47
Little Liu, 133
longevity, 81; pollution and, 98, 101
Lu Hanlong, 169n4

mahjong, 133
Malinowski, Branislaw, 168n3
Mandarin, standard, xiii; Romanizing, 167n1
mandate of heaven, 122, 156